BEYOND THE PILLARS OF HERCULES

Beyond the Pillars of Hercules

Atlantis and Tyrus in Plato's Writings, Biblical
Verses, and the Works of Helena Blavatsky,
Edgar Cayce, and Ruth Montgomery

David Hershiser

Sun on Earth™ Books
Heathsville, Virginia

Published by Sun on Earth™ Books

www.sunonearth.com

Illustrations by the author.

Publisher's Cataloging-in-Publication Data
Hershiser, David.
    Beyond the pillars of hercules / David Hershiser.— Ist ed.
      p. cm.
    ISBN: 978-1-883378-64-6
        1. Atlantis
        I. Title.

GN751.B49 2014
001.94—dc23

Library of Congress Control Number: 2014934728

ISBN: 978-1-883378-64-6

Mays, James L., editor. *Harper's Bible Commentary*. San Francisco: Harper and Row, 1988.

McKenzie, S.J., John L. *Dictionary of the Bible*. New York: Macmillan Publishing Co., Inc., 1965.

The New Interpreter's Bible. Nashville, TN: Abingdon Press, 2001.

The New Encyclopedia Britannica, 15th Edition. Chicago: Encyclopedia Britannica, Inc., 2005.

Smith, LL.D., William. *Smith's Bible Dictionary*. New York: Pyramid Books, 1967.

Articles

Kellner, Mark. "The 400-Year Reign of the King James Bible." *Washington Times*, January 3, 2011.

Langenbrunner, Norman. "How to Understand the Bible: Examining the Tools of Today's Scripture Scholars." *Publication #0382, St. Anthony Messenger Press*, 1982.

"Researchers 'find' Atlantis in Spain." *Venture Inward Magazine*, Edgar Cayce's Association for Research and Enlightenment: July-August-September 2011.

Online Resources

Atsma, Aaron J. "Island Erytheia." *The Theoi Project: Greek Mythology*. Accessed December 15, 2013. http://www.theoi.com/Kosmos/Erytheia.html.

Blavatsky, H.P. "The Secret Doctrine, vol.2." *Theosophical University Press Online Edition*. Accessed December 15, 2013. http://www.theosociety.org/pasadena/sd/sd-hp.htm.

Charles, R. H. "The Book of Jubilees." *The Apocrypha and Pseudepigrapha of the Old Testament*. Accessed December 15, 2013. http://www.pseudepigrapha.com/jubilees/7.htm.

"Gentile." *Wikipedia*. Accessed December 15, 2013. http://wikipedia.org/wiki/Gentile.

Herodotus. "The Histories, Book IV." *Ancient History Sourcebook: Accounts of Ancient Mauretania, c.430 BCE – 550 CE*. Accessed December 15, 2013. http://www.fordham.edu/halsall/ancient/anc-nafrica.asp.

Strabo. "Geography, vol. XVII.iii." *Ancient History Sourcebook: Accounts of Ancient Mauretania, c.430 BCE – 550 CE*. Accessed December 15, 2013. http://www.rbedrosian.com/Classic/strabo17d.htm.

The Unbound Bible. Accessed December 15, 2013. http://unbound.biola.edu.

CONTENTS

TIMELINE

10000 B.C.	End of the Pleistocene Epoch and the last Ice Age
9600 B.C.	Approximate date for the destruction of Atlantis according to Plato
7000 B.C.	Beginning of civilization in the Middle East
2100 B.C.	Phoenician cities of Tyre and Sidon known to be in existence
742 B.C.	Beginning of Isaiah's ministry
626 B.C.	Beginning of Jeremiah's ministry
593 B.C.	Beginning of Ezekiel's ministry
587 B.C.	Beginning of Nebuchadnezzar's unsuccessful fourteen-year siege of Tyre
427 B.C.	Approximate birth year of Plato
332 B.C.	Phoenician city of Tyre conquered by Alexander
A.D. 405	The Vulgate – Jerome's Latin translation of the Bible completed
A.D. 1611	King James Bible published
A.D. 1871	Edgar Cayce's birth year
A.D. 1882	Ignatius Donnelly's *Atlantis: the Antediluvian World* published
A.D. 1888	Helena Blavatsky's *The Secret Doctrine* published

INTRODUCTION

In his two dialogues—the *Timaeus* and *Critias*—Plato used the phrase "the Pillars of Hercules" to refer to what we know today as the Strait of Gibraltar. His discussion of Atlantis in the Timaeus is relatively brief, but the Critias contains a longer and much more detailed account of Atlantis and its history.

According to Plato, Atlantis was an island in the Atlantic Ocean located beyond the Pillars of Hercules. It was larger than "Libya and Asia put together," by which he probably meant North Africa plus the Middle East and India. Its easternmost portion reached to a point just outside the Strait of Gibraltar and faced the region of southwest Spain. Plato mentioned in the Timaeus the existence of other islands in the ocean near Atlantis, and that from these islands, travelers could reach "the opposite continent." Although he didn't mention the name of this continent, many authors believe he was referring to America. If true, this would have been remarkable, since knowledge of the American continent was supposedly unknown in Plato's time. Others believe that Atlantis consisted of a series of islands stretching across the Atlantic Ocean— from the entrance of the Mediterranean Sea in the east, to the Caribbean Islands in the west.

The existence of Atlantis has been the subject of debate ever since Plato discussed it in his dialogues.

Many critics of a historical Atlantis believe that Plato simply invented the island and used it in the Timaeus and Critias to illustrate his vision of an ideal society.

Countless books have been written about Atlantis, and many theories have been proposed for its location. These range from Antarctica in the south to Scandinavia in the north, and from the Caribbean in the west to the Mediterranean in the east.

The modern consensus for the location of Atlantis centers on the Greek island of Santorini. Formerly called Thera, it was nearly destroyed in 1500 B.C. by a massive volcanic eruption. Many people believe that Plato based his account of Atlantis on this event. The tsunami generated from the explosion also destroyed the Minoan civilization on Crete. The date of Thera's destruction in 1500 B.C., however, was much more recent than the date Plato gave for the destruction of Atlantis, which he said was over 9,000 years before his own time, or about 9600 B.C.

Plato's dialogue is regarded as the first account of Atlantis in western literature. If Atlantis really existed, then surely there must have been other books written about it in ancient times that are no longer available to us.

The Great Library of Alexandria, once a repository of knowledge for the ancient world, was thought to contain supportive evidence for Atlantis. The destruction of the library early in the first millennium

A.D., and the records and scrolls contained there, was certainly a great tragedy. In his popular 1982 book and TV documentary *Cosmos,* Carl Sagan mentioned a three-volume work written by the Babylonian priest Berossus that was contained in the library's collection. The first volume covered the time from the Creation to the Flood, a period Berossus reckoned lasted 432,000 years. It isn't known exactly what was in his book, but many early writers cited Berossus in their own works. If Atlantis actually existed, and was as powerful and far-reaching as Plato described it, Berossus would likely have mentioned it in his book.

Could there be another source of the Atlantis legend—one that predates Plato, is familiar to us all, and is widely available even now? According to the nineteenth-century theosophist Helena Blavatsky, an account of Atlantis can be found in the Bible. She stated that Ezekiel's prophecy about the destruction of Tyrus was really about Atlantis. The city is mentioned many times in the Old Testament, especially in Ezekiel's long account of the city (Ezekiel 26–28), but also in the prophetic books of Isaiah, Jeremiah, and Zechariah.

Blavatsky's claim seems preposterous in light of the accepted scholarship on the subject. There are no arguments among serious scholars that Tyrus was anything other than a prosperous Phoenician city on the coast of what is now Lebanon. Usually referred to as Tyre, the city has a long, well-documented history.

If the King James Version of the Old Testament is studied carefully, however, and the passages concerning Tyrus are considered literally, many questions arise about this supposed Phoenician city. Many of these passages don't accurately reflect the historical record of the Phoenician city. They often seem to describe a different city altogether—a city that was located somewhere else, and one that was much wealthier, more powerful, more warlike, and one which Ezekiel prophesied would ultimately disappear beneath the sea in a violent cataclysm. Many of the things Ezekiel said about Tyrus, at least in the King

James Version (KJV), mirror what Plato said about Atlantis. Much of the material from the KJV that seems to be about Atlantis, however, has been redacted from modern translations of the Bible, or altered to the extent that any connection to Atlantis has been rendered all but invisible.

1

Blavatsky and Tyrus

Helena Blavatsky

Helena Petrovna Blavatsky, usually referred to as Madame Blavatsky, was born in Russia in 1831 and came to New York City in 1873. A controversial figure, she was widely traveled, well read, a prolific writer, and said to have a magnetic personality. While living in New York, she co-founded the Theosophical Society in 1875 with Henry Olcutt. Shortly thereafter, even though married to someone else at the time, Blavatsky left New York with Olcutt and traveled to India, where they remained for six years. She eventually returned to London, where she died in 1891.

Although sometimes regarded as a fraud and a plagiarist, Blavatsky nevertheless did much to further the understanding of Eastern religious, philosophical, and occult concepts throughout the Western world. She remains a leading name in the New Age movement.

According to Blavatsky, Theosophy was "the archaic wisdom-religion, the esoteric doctrine, once known in every ancient country having claims to civilization." Her philosophy is summarized in her best-known work, *The Secret Doctrine*, published in 1888. This book is regarded as the primary source for an understanding of modern theosophy. She said the material in the book was divinely revealed to her and was based on a Tibetan manuscript called the *Stanzas of Dzyan*. It is in The Secret Doctrine that her speculations on Atlantis and Tyrus can be found.

"The origin of the 'prince of Tyrus'," she wrote, "is to be traced to, and sought in the 'divine Dynasties' of the iniquitous Atlanteans, the Great Sorcerers." She said Ezekiel's prophecy for the destruction of Tyrus was not prophecy but simply a reminder of the fate of the Atlanteans, the "Giants in the Earth."

She added that chapters twenty-eight and twenty-nine of Ezekiel actually refer to Atlantis and not to Babylon, Assyria, or Egypt, because these places fell into ruins on the surface, of which evidence still exists, unlike Atlantis, of which no evidence remains.

Was Ezekiel's account of Tyrus really about Atlantis, as Blavatsky claimed? In an article for the magazine *Atlantis Rising*, Frank Joseph wrote that a belief in Atlantis was at one time considered heretical because it was not mentioned in the Bible. Modern theologians certainly do not regard Atlantis as an explanation for any of the events described in the Old Testament. A careful study of the *New Interpreter's Bible* (NIB), a standard reference work in the modern interpretive literature, will result in a fruitless search for any mention of Atlantis.

According to Robert Wilson in *Harper's Bible Commentary*, Ezekiel has always been a difficult book to understand, for lay readers and scholars alike. Ezekiel's ministry took place between 593 B.C. and 571 B.C., but it is believed the book was written at a much later date, probably around 230 B.C. The Babylonian

Talmud reported that some rabbis even recommended banning the reading of Ezekiel for anyone under the age of thirty, for fear it would lead to dangerous mystical speculations, or even destroy the interpreter who looked too deeply into its mysteries. St. Jerome himself, who began working on the Vulgate Bible in A.D. 382, had difficulties with Ezekiel and apologized in his commentaries for his inability to make sense of many of the obscure passages. Ezekiel remains a puzzle even today, and it is unlikely that scholars will ever agree on the meaning of this confusing book.

Although Blavatsky wrote that Ezekiel's account of Tyrus was really about Atlantis, she gave only a few reasons for her claim. Many of the passages about

Tyrus that are found in Ezekiel, as well as in the other prophetic books of the Old Testament, can be interpreted to support the idea of a biblical Atlantis. Many of these passages parallel very closely the things Plato said about Atlantis in his dialogues. In Ezekiel, where the longest account of Tyrus can be found, one passage in particular is critical to the argument for a biblical Atlantis.

2

The Entry of the Sea

View of Tyrus beyond Gibraltar

In the King James Bible, Ezekiel describes the location of Tyrus.

And say unto Tyrus, O thou that art situate at the entry of the sea, which is a merchant of the people for many isles, Thus saith the Lord God; O Tyrus, thou hast said, I am of perfect beauty. — Ezekiel 27:3

The meaning of the phrase "the entry of the sea" in this passage seems perfectly clear. The sea mentioned here is the Mediterranean Sea, once known as the Great Sea, and the entry of the Mediterranean Sea is the Strait of Gibraltar, or as Plato called it, the Pillars of Hercules. This passage simply says that Tyrus was located somewhere in the vicinity of the Strait of

Gibraltar. Plato said precisely the same thing about Atlantis—that the easternmost portion of the island was located just outside the Strait of Gibraltar, and that it faced the southwestern portion of today's Spain. Plato referred to this region of Spain as Gadira, which meant the area near present-day Cádiz.

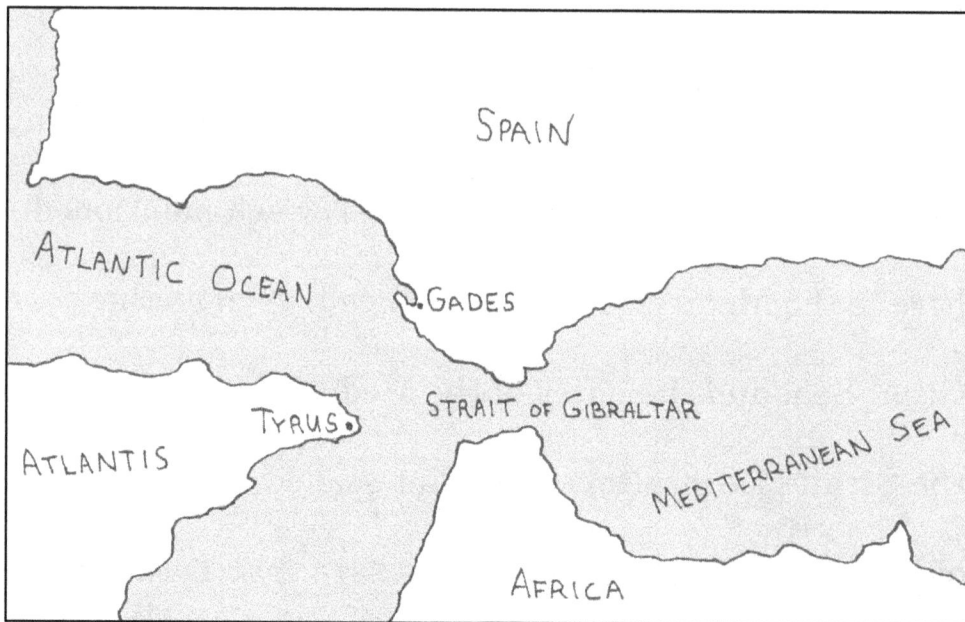

Concept of the region around Gibraltar, ca. 10000 B.C.

Another passage, this time from Jeremiah,

describes the location of Tyrus in much the same way.

And all the kings of Tyrus, and all the kings of Zidon, and
the kings of the isles which are beyond the sea.
— Jeremiah 25:22

At first glance, the meaning of this passage seems

to be somewhat ambiguous. Does Jeremiah say that

Tyrus and Zidon *and* the isles are all located beyond

the sea? Or does he say that only the isles themselves

are beyond the sea, while the cities of Tyrus and Zidon

are not? The second interpretation seems to be the one

favored by scholars, but it seems reasonable to assume

that Jeremiah meant all of them—Tyrus and Zidon and the isles—were located beyond the sea. The phrase "beyond the sea" obviously seems to mean beyond the Mediterranean Sea and the Strait of Gibraltar, and somewhere in the Atlantic Ocean.

Do these two passages from Ezekiel and Jeremiah really describe an island in the Atlantic Ocean near Gibraltar? Do they validate Blavatsky's claim that Ezekiel's account of Tyrus was really about Atlantis? Maybe so, but not according to the commentary from the *New Interpreter's Bible* (NIB). The explanation given in the NIB is based on the assumption that the cities of Tyrus and Zidon were the Phoenician cities of Tyre and Sidon. The different spellings—Tyrus and

Tyre, and Zidon and Sidon—are used interchangeably throughout the King James Bible, but theologians always refer to Tyrus as Tyre, and Zidon as Sidon when they are the subject of commentary and analysis.

The Phoenican City of Tyre

Tyre, now called Sur, was a wealthy Phoenician city-state located on a small island 800 yards off the coast of present-day Lebanon. The island is now connected to the mainland by an isthmus, and lies about fifty miles south of Beirut. Alexander the Great successfully conquered the city in 332 B.C., in part because he built a causeway connecting the island city to the mainland. The causeway remained in place, and over the years silt has filled in the harbors, giving the city its present configuration. Several hundred years before Alexander's conquest of the city, and while it was still an island, Tyre endured a long, unsuccessful siege by Nebuchadnezzar, from 587 B.C. until 574 B.C., which was during the time of Ezekiel and the Babylonian exile of the Jews. The city has a long, well-documented history, and is generally assumed to have been in existence since at least 2100 B.C.

More recent translations of the Bible gradually embrace the assumption that Tyrus and Tyre were the same city. Although the descriptions of Tyrus found in the King James Bible don't quite fit the known historical circumstances of the Phoenician city, it is obvious that over the course of successive translations, they have been rewritten to do just that.

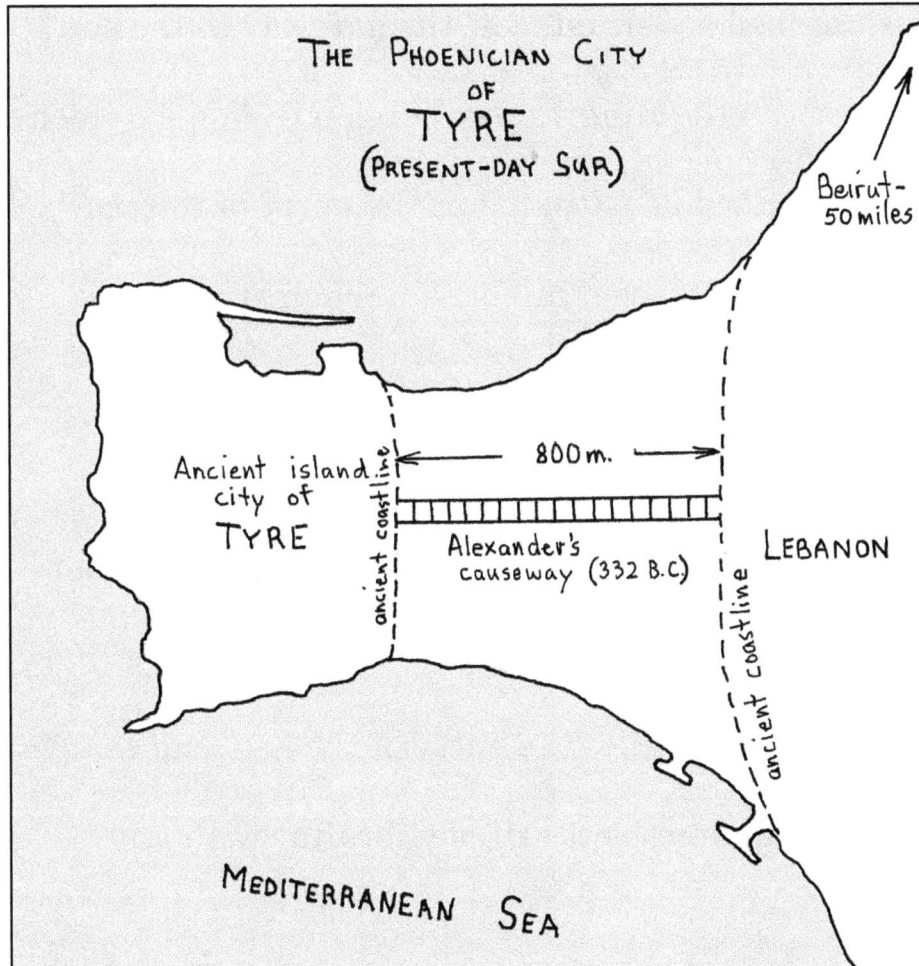

The Phoenician city of Tyre, as it looks today

In the *New International Version (NIV)* of the Bible, published in 1978, for example, the location of Tyrus is now described as "situated at the gateway to the sea," in contrast to the King James Version (KJV), in which it was described as "situated at the entry of the sea." This change may seem minor, but when studied carefully, the newer translation more accurately describes the location of the Phoenician city of Tyre. The newer phrase, "the gateway to the sea," is rather ambiguous, and could mean anywhere in the Mediterranean, while "the entry of the sea" can really only be applied to the Strait of Gibraltar with any accuracy.

Another example from *The Good News Bible* (1976) shows an even greater change to the original phrase, from "the entry of the sea" to "Tyre, that city which stands at the edge of the sea." The transformation is complete: Tyrus has now become the Phoenician city of Tyre.

Tyrus and Tarshish

The Book of Isaiah also contains an account of Tyrus. This account is substantially shorter than the one found in Ezekiel, and refers to the city as Tyre, rather than Tyrus. It describes the city in the past tense, as already having been destroyed. This is in contrast to the prophecy found in Ezekiel, where the destruction of

Tyrus has not yet occurred. One passage in particular from the King James Version of Isaiah illustrates the problem with the scholarly assumption that Tyrus and Tyre were the same city.

Pass ye over to Tarshish; howl, ye inhabitants of the isle.
— Isaiah 23:6

The isle in this passage refers to Tyrus, while the city of Tarshish, usually associated with the classical Roman city Tartessos, is generally believed to have been situated near Cádiz in southwestern Spain. Upon reading the passage, it might be assumed that Tyrus and Tarshish were located relatively near to each other. But if it were true that Tyrus and Tyre were the same Phoenician city, and located on the coast of Lebanon, why would the inhabitants be told to "pass over" to Tarshish, which was located on the opposite side of the Mediterranean Sea, through the Strait of Gibraltar, and on the Atlantic coast of Spain—a distance of approximately 2,300 miles to the west? The passage makes more sense if Tyrus had been an island in the vicinity of Gibraltar. It seems logical to assume that the inhabitants of an island facing destruction would flee to the nearest mainland, rather than set sail for a destination thousands of miles away. If Isaiah was actually speaking about the Phoenician city of Tyre here, why not just cross over to the mainland of Lebanon, a mere 800 yards away? If this happened to be hostile territory, surely there were other safe havens in the eastern Mediterranean much closer than Tarshish.

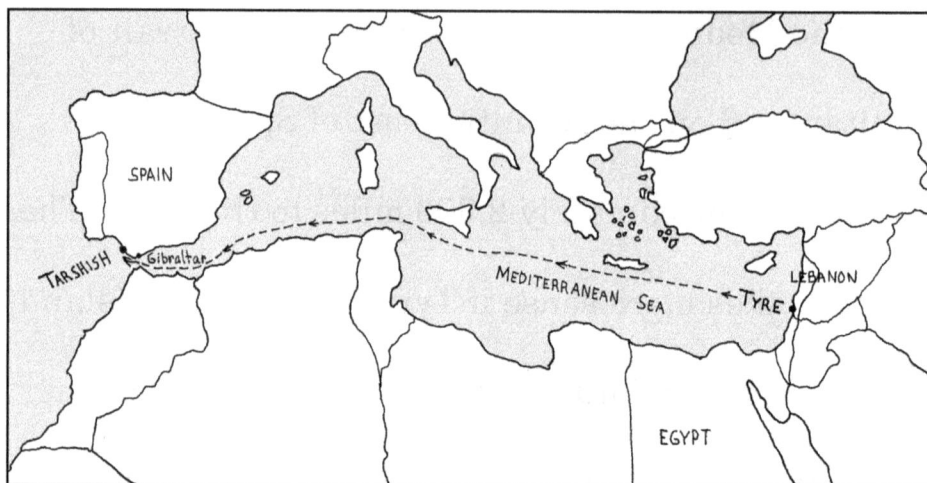

Tyre to Tarshish, a distance of 2,300 miles

There is one final observation to be made about the location of Tyrus. The city is described many times by Ezekiel as lying in the "midst of the seas."

The following examples are typical and are from the King James Bible.

Thy borders are in the midst of the seas.
— Ezekiel 27:4

What city is like Tyrus, like the destroyed in the midst of the sea?
— Ezekiel 27:32

It shall be a place for the spreading of nets in the midst of the sea.
— Ezekiel 26:5

These descriptions of Tyrus seem somewhat exaggerated when compared to the actual geographical and physical circumstances of the Phoenician city of Tyre, which was located on a small island in relatively shallow coastal waters only a few hundred yards from shore. The Phoenician city can hardly be accurately described as lying in the midst of the seas. But if Blavatsky's claim is correct, then the description does seem appropriate for an island in the Atlantic Ocean, such as Atlantis.

It is instructive once again to compare the KJV passages with their newer translations. The first passage, "Thy borders are in the midst of the seas," becomes, in the New International Version (1978), "Your domain was on the high seas." At first glance, the two seem close enough, but after careful study, it can be seen that they are actually quite different. "Borders in the midst of the seas" describes the city's physical location, whereas "Domain on the high seas" more accurately describes the city's political or economic sphere of influence. The 1978 translation, in this case, is another example of the scholarly assumption that Tyrus and Tyre refer to the same place—a successful Phoenician seaport that conducted most of its commerce by sea.

3

The Ship of Tyre

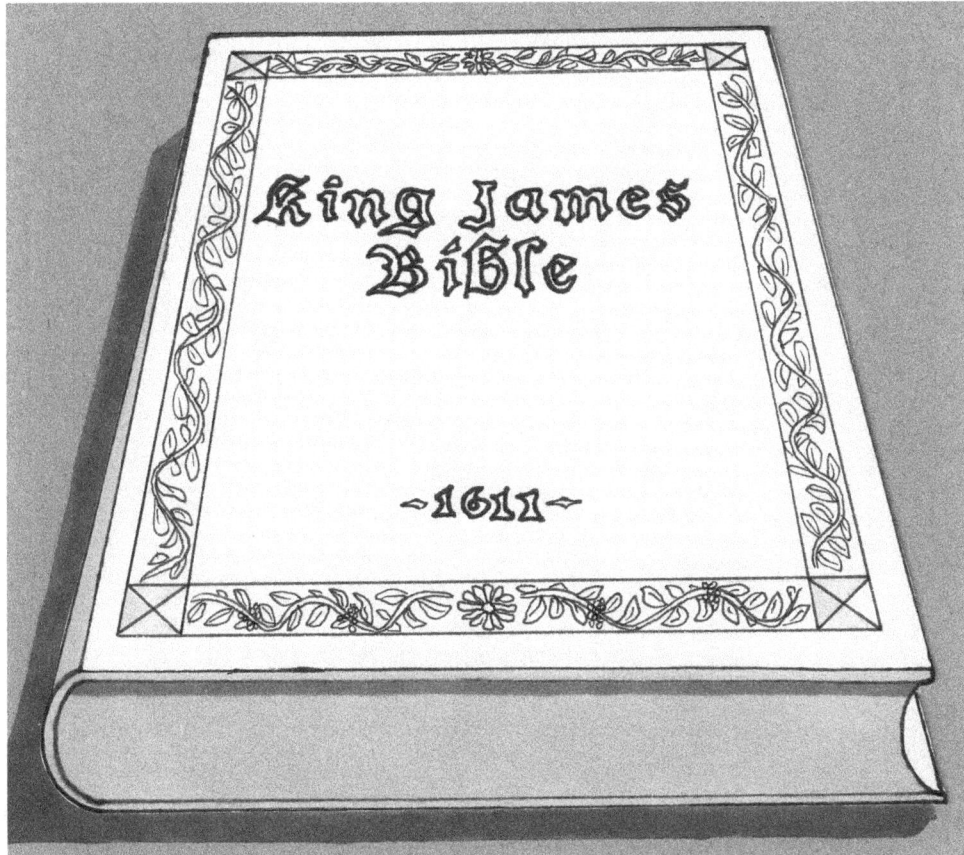

In The Secret Doctrine, Blavatsky wrote the following about Ezekiel's account of Tyrus. "There is no metaphor here, except in the preconceived ideas of our theologians, perhaps."

According to the *New Interpreter's Bible*, Ezekiel was considered to be a master of metaphor, and his descriptions of Tyrus are a testament to his literary prowess. The NIB assumes that Ezekiel must have used maritime metaphors to emphasize the commerce and trade in which Tyre engaged with the ancient world. This view probably stems from Ezekiel's descriptions of Tyrus from the King James Bible, which, as noted in the previous chapter, do not closely reflect the historical circumstances of the Phoenician city of Tyre. The very idea that Tyrus and Tyre may have been different cities seems out of the realm of consideration for most scholars.

The NIB continues the metaphor by regarding the city—with its far-reaching commercial activities involving the shipping of a variety of luxury goods to and from a large number of trading partners—as a "ship of dreams." Later, the metaphor concludes when Ezekiel prophesies the sinking of the "ship of Tyre" due to its arrogance and greed, followed by the anguished responses from the city's trading partners and fellow kingdoms when they hear of the catastrophe of the "shipwreck."

Scholars conclude that the story of Tyre, and its metaphorical portrayal as a ship must be the correct interpretation because the Phoenician city of Tyre exists to this day. Although it waned in power and influence long ago, it never disappeared beneath the sea as Ezekiel's prophecy foretold. His prophecy of destruction went unfulfilled. How else, then, could the story be interpreted, other than as a metaphor?

The explanations found in the *New Interpreter's Bible* do not agree with Blavatsky's interpretation of Ezekiel. In fact, her claim runs counter to the conclusions reached by a formidable array of experts on the subject. Published in 2001, the NIB was written by 900 scholars, all experts in their fields and representing many different theological perspectives. The NIB was written for use by the clergy rather than the casual reader. It replaced the *Interpreter's Bible (IB)*, which was published in 1952. Both editions are multi-volume reference works that provide line-by-line analysis of each book of the Bible. One important difference between the two is that the earlier edition has the text of the King James Bible set side-by-side with the text of the RSV (Revised Standard Version, 1952) for the purposes of comparison, while the newer edition uses the texts of the NIV (New International Version, 1978) and the NRSV (New Revised Standard Version, 1989) for comparison. The change in texts is significant because newer Bible translations incorporate the assumption that Tyrus and Tyre were the same city, thus making it difficult to see any connection with Atlantis. This connection is much more evident in the King James Version, published in 1611.

When understood literally, the examples from the King James Bible discussed so far seem at odds with modern assumptions about Tyrus that are embedded in the commentaries from the NIB and other reference sources. There are many more passages in the King James Bible that can be interpreted in a similar fashion.

4

The Wealth of Atlantis and Tyrus

Atlantis

In addition to their apparently identical locations, Tyrus and Atlantis were both described as fabulously wealthy. Plato wrote of the incomparable wealth of Atlantis in the Critias, providing many examples of the manner in which it was displayed.

...their wealth was greater than that possessed by any previous dynasty of kings or likely to be accumulated by any later.

Plato described the "outlandish" or "barbaric" splendor of the Temple of Poseidon.

He went into great detail explaining the various
attributes of this amazing structure. The outside of the
temple was covered with silver, and the figures on the
pediment were covered with gold. On the inside, the
ceiling was also covered with gold and silver, and
picked out with ivory. The rest of the interior was
covered with the mysterious metal orichalc, which

Plato said was second only to gold in value. It is not known with any certainty what orichalc was, but it has been suggested that it was an alloy of copper and gold. The dimensions of the temple were extraordinary—300 feet wide, a stade in length (about 600 feet), and proportionate in height.

Comparison of the Temple of Poseidon (left) with the Parthenon and the Great Pyramid of Giza

These measurements are three times the width, the length, and the height of the Parthenon. In terms of volume, twenty-seven Parthenons could have fit inside the Temple of Poseidon. It covered an area equal to four football fields. The focal point of the temple was a gold-plated statue of Poseidon surrounded by 100 Nereids, or sea nymphs, that reached to the ceiling. Plato's description of the temple is difficult to believe. It would have required huge amounts of gold and silver to decorate such an enormous structure.

Consider the one remaining wonder of the ancient world—the Great Pyramid of Giza. What if it also, like the Temple of Poseidon, existed only as a reference in an ancient text? Would it be regarded as a myth?

Would we doubt that it ever existed? The pyramid, 755 feet on each of its four sides and 480 feet tall, is much larger than the temple Plato described. In the book *Pyramid Odyssey* (1978), William Fix wrote that this massive granite and marble structure was once encased in twenty-one acres of polished white limestone. Built with the precision of a Swiss watch, the pyramid has been the subject of many books filled with explanations and speculations on its true meaning and purpose.

Another example of the vast wealth of Atlantis can be found in Plato's description of the royal residence on the island itself. This city consisted of a small central island or acropolis five stades (3,000 feet) in diameter.

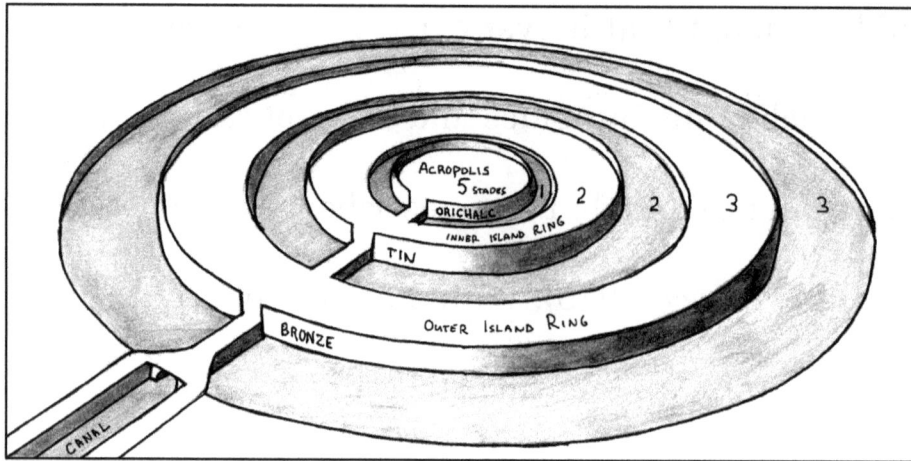

The Temple of Poseidon was located there. Surrounding the acropolis were three concentric rings of water and two of land. Each ring of land was protected by a wall, which Plato described as "of some height above sea level." Each wall, in turn, was covered with a thin sheet of metal.

The wall encircling the acropolis was covered with orichalc. If this wall had been twenty feet high, for example, then, based on Plato's dimensions, the surface area of the wall would have equaled 4.32 acres, completely covered with a sheet of the precious metal orichalc.[1]

Similarly, the wall around the next island ring was coated with tin, and based on Plato's measurements, it would have had a surface area of 9.5 acres. What was the source of such a large amount of tin? Many scholars associate Tarshish with the city of Tartessos, which was located near Cádiz in the area of southwestern Spain, and known to be a source of tin during the Bronze Age. Phillip M. Freeman, in *Celtic from the West* (2010), notes that the fourth-century-B.C. Greek

---

[1] The acreage of the wall is calculated in the following manner: 5 stades (or 3,000 feet) times pi, times 20—with the result divided by 43,560.

historian Ephorus described Tartessos as very prosperous and "with much tin carried by the river." Although the European Bronze Age (tin being an important component of bronze) began around 2000 B.C.—long after Plato's date for the destruction of Atlantis in approximately 9600 B.C.—the following passage from Ezekiel describes Tarshish as a supplier of tin for Tyrus.

> *Tarshish was thy merchant by reason of the multitude of all kinds of riches; with silver, iron, tin, and lead, they traded in thy fairs.* — Ezekiel 27:12

Does this passage show a link between Tyrus and Atlantis? It clearly states that Tyrus received tin from Tarshish. If Tyrus was a part of Atlantis, as Blavatsky said, and thus in existence long before the European Bronze Age, does this passage suggest that Tarshish was the source of the tin for the walls of Atlantis that were described by Plato in the Critias?

Finally, Plato's description of the outer wall of Atlantis is astonishing. The wall around the perimeter of the outer island ring was covered with bronze. Based on the dimensions Plato gave in the Critias, it would have had a surface area of 18.1 acres. Bronze is an alloy of copper and tin. What was the source of the copper needed for the manufacture of such an enormous amount of bronze?

Perhaps it came from the copper mines located on the upper peninsula of Michigan on the shores of Lake Superior. Evidence of mining in this region has been dated to as early as 5000 B.C., and, in one case, to 7000 B.C. An enormous amount of copper was taken from the ground here in ancient times. Some estimates range as high as 1.5 billion pounds, an amount experts regard as excessive, but more reasonable estimates still yield a staggering amount of copper. It is assumed Native Americans mined this copper, although nobody really knows. These mysterious early miners disappeared and left little or no evidence of their existence. Not even burial grounds have been found.

An even more puzzling question is what happened to all of this copper. No developed cultures existed at that time in America to use it, let alone a Bronze Age culture, and only a small percentage of the copper has been accounted for in America. Some authors have speculated that explorers from Atlantis found these deposits of copper and mined them for their own ends. With its purported maritime capabilities and extensive trading empire, maybe a large portion of the copper was transported to the capital city of Atlantis and used to gild the walls of this fabulous city.

Ezekiel likewise made numerous references to the wealth of Tyrus.

With thy wisdom and with thine understanding thou hast gotten thee riches, and hast gotten gold and silver into thy treasures: By thy great wisdom and by thy traffick hast thou increased thy riches, and thine heart is lifted up because of thy riches...

— Ezekiel 28:4–5

The *New Interpreter's Bible* states that Ezekiel wrote these passages to illustrate the lesson that the pursuit of excessive wealth ultimately leads to the destruction of society. Although his prophecy never came to pass, and the Phoenician city still exists, the same cannot be said of Atlantis.

While these passages might be regarded as examples of Ezekiel's mastery of metaphor, the following description of Tyrus from Zechariah, one of the Minor Prophets, seems exaggerated even when regarded as a metaphor.

And Tyrus did build herself a strong hold,
And heaped up silver as the dust,
And fine gold as the mire of the streets.

— Zechariah 9:3

Although it is certainly true that the Phoenicians were among the wealthiest communities of the ancient world, this verse seems better suited as a description for Atlantis than for the Phoenician city. Whether or not it was simply poetic license on the part of Zechariah, and a supposed metaphorical description of the Phoenician city of Tyre similar to the descriptions provided by Ezekiel, the passage would serve very nicely in Plato's Critias, where Atlantis was described as wealthier than any place before or since.

5

Gadirus

Plato

Plato gave a very detailed account of the history and founding of Atlantis in the Critias, as well as a thorough description of the city itself. He wrote that when the gods divided up the earth between them, Poseidon was allotted the island of Atlantis.

At a later time, Poseidon became attracted to an earth-born woman named Cleito, and she bore him five sets of male twins. The island was divided into ten parts, and each twin was allotted one of the districts. The first-born son, Atlas, was made King of all the others, and the whole island and the surrounding ocean were named after him. His twin brother Gadirus was given the easternmost part of Atlantis near the

Pillars of Hercules. This part of the island faced the district Plato called Gadira, which we know today as the region of southwestern Spain near Cádiz.

The names of many ancient cities in Spain, as well as throughout the Mediterranean, seem to have a common origin in the name of Gadirus. Many Greek writers such as Strabo, Herodotus, and Pliny, as well as the nineteenth-century writer Ignatius Donnelly, have either implied or stated directly that the source of many of the ancient civilizations found inside the Strait of Gibraltar and around the shores of the Mediterranean Sea was Atlantis. The Tyrians, the Etruscans, the Carthaginians, and the Phoenicians fall into this category and have all been linked to Atlantis.

Cádiz

Cádiz is a port city on the coast of southwestern Spain, and the capital of one of the provinces that make up Andalusia. It is believed to be the oldest continuously inhabited city in Spain, and perhaps in all of southwestern Europe. Cádiz lies about 100 miles north of Gibraltar and is believed to have been founded by the Phoenicians in 1104 B.C.

Plato wrote in the Critias that the region of Spain surrounding Cádiz was named after Atlas's twin, Gadirus, but it isn't clear from his explanation just how the name of the city originated. Later, in 22 B.C., the Greek writer Strabo gave a better explanation of the origin of the word Cádiz in his book *Geography*. Strabo explained that the word was originally spelled *Gades*, which in turn was derived from the name *Gadirus*.

If the easternmost portion of Atlantis was located near the coast of southwest Spain, as Plato said, it is possible that a city in this region might also bear the name of the Atlantean King presiding nearby.

Recent Archaeological Finds in Cádiz

Some recent discoveries in the Cádiz area may prove to be of interest. Beginning in 2009, an international and American research team from the University of Hartford analyzed satellite pictures of ruins beneath the marshlands of Doña Ana Park, just north of Cádiz. Head researcher Richard Freund believes they may have discovered Atlantis. Using deep-ground radar, digital mapping, and underwater technology to study the site, they've found a series of ringed cities that resemble the city of Atlantis Plato described in the Critias.

If Atlantis was located in the Atlantic Ocean, where Plato said it was, it seems unlikely that these recent finds are remnants of the lost city. Their proximity to the same area, however, might reveal new information about Atlantis.

Mt. Atlas and Gadirus

In *Geography*, Strabo wrote that Mt. Atlas was once called Mt. Dyris by the barbarians, by which he meant the name of Mt. Dyris came from the native people of the area surrounding the mountain. Based on the obvious similarity between Gadirus and Dyris, it isn't much of a stretch to assume that the latter is derived from the former. This is particularly true when it is considered that both Herodotus and Strabo said Mt. Dyris was located in the same region of the world where Plato said the part of Atlantis ruled by Gadirus was located.

Both Strabo and Herodotus mentioned Mt. Atlas in their writings, and both said it was located somewhere in northwestern Africa in the vicinity of Gibraltar. Based on their writings, however, the exact location is difficult to pinpoint. Strabo said it was just past the

Pillars of Hercules near what is now Tangiers. Herodotus wrote that Mt. Atlas was so lofty, its top couldn't be seen, and the natives of the area, who called themselves Atlantes, named it "the Pillar of Heaven."

There seem to be two likely candidates for Mt. Atlas in this part of the world. The first, Mt. Toubkal, is the highest peak in the Atlas Mountain Range. It reaches 13,671 feet in height and lies some thirty-nine miles south of Marrakesh in southwest Morocco.

The second, and perhaps more likely candidate, is the volcanic cone called Mt. Teide. It lies on the island of Tenerife in the Canaries. This Spanish island chain forms the westward extension of the Atlas Mountain range.

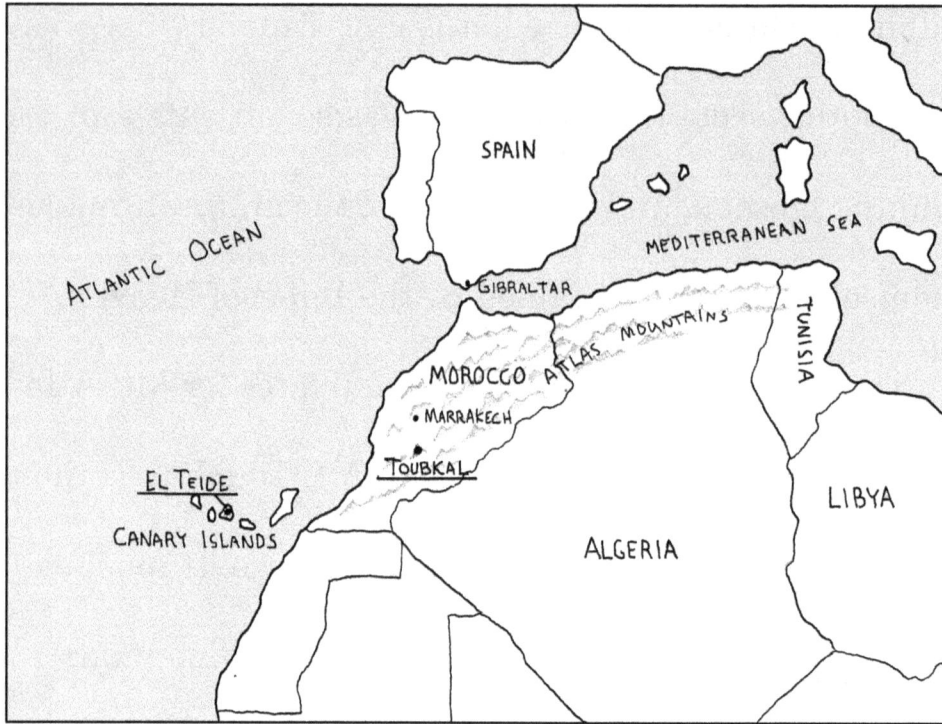

At 12,198 feet, Mt. Teide is the highest point in
Spain, and also the highest point above sea level
among the islands of the Atlantic. When measured
from its base on the ocean floor, it is the third largest
volcano in the world. Only the Hawaiian volcanoes,

Mauna Kea and Mauna Loa, are higher. Local legend has it that the Canary Islands are the uppermost peaks of Atlantis and are the only portion of the sunken continent still remaining above sea level.

In his book *Earth in Upheaval*, Immanuel Velikovsky mentioned the French savant A. Berthelot, who believed the Atlas Mountains were torn apart in a great upheaval that resulted in the Canary Islands being cut off from the African continent. According to Velikovsky, this upheaval also resulted in the emptying of Lake Triton, leaving behind what is known today as the Sahara Desert.

Helena Blavatsky also discussed Mt. Atlas in *The Secret Doctrine*, remarking that, according to legend, it was once three times its present height. She doesn't reveal the source of the legend, but it coincides with the theory that when Atlantis sank beneath the sea, perhaps due to the collapse of the earth's crust under the mid-Atlantic ridge, Mt. Atlas and the surrounding territory became lower in elevation as well.

If the apparent connection between *Dyris* and *Gadirus* is true, then it seems reasonable to conclude that *Tyrus* is also derived from *Gadirus*. Apart from the obvious similarity between the two words, when it is also pointed out that Ezekiel's description for the location of Tyrus is identical to Plato's location for the easternmost portion of Atlantis, the link between the words Tyrus and Gadirus seems obvious.

Tyrrhenia

Plato wrote in the Timaeus that Atlantis controlled the Mediterranean Sea from northwest Africa to Egypt, and Europe as far as Tyrrhenia.

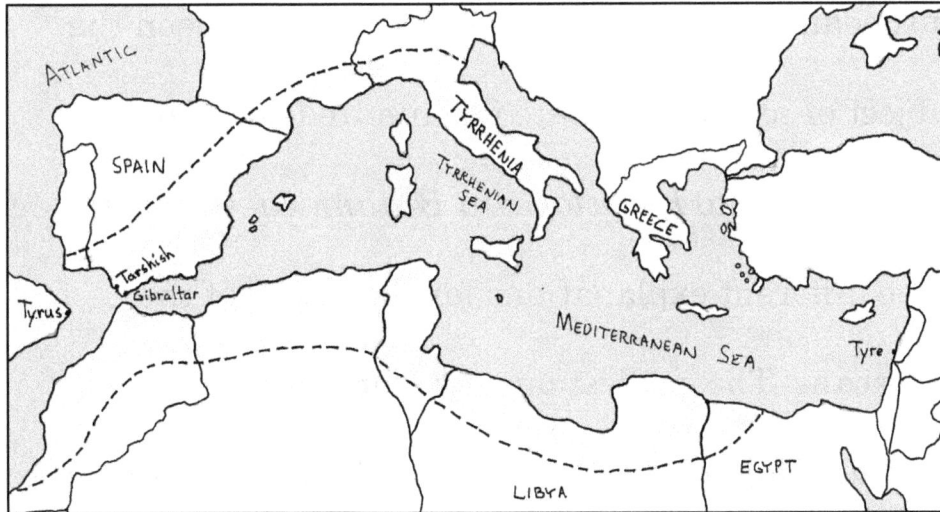

Tyrrhenia is the name Plato and the ancient Greeks used to refer to ancient Italy. The people were called the Tyrrhenoi, and the part of the Mediterranean Sea adjacent to the west coast of Italy still goes by the name of the Tyrrhenian Sea. In English, the Tyrrhenoi are known more familiarly as the Etruscans, a Latin word which stems from the name of their country, Tuscia or Ertruria. The Etruscans spoke a non-Indo-European language, and their origins have been the subject of scholarly debate for hundreds of years.

According to *Encyclopedia Britannica*, there are three different explanations for the origin of the Etruscans. The earliest one, by Herodotus, claimed that they migrated from the Black Sea region after the Trojan War. Their leader, Tyrsenos, gave his name to the whole race.

Later, the Greek writer Dionysius stated that the Etruscan civilization began from local origins. The most recent theory, from the nineteenth century, holds

that the Etruscans migrated to Italy from the north.

The *Encyclopedia Britannica* concludes that none of

these theories offer any firm proof of the origins of the

Etruscans. There is, however, a radically different

theory popularized by the nineteenth-century writer

and congressman Ignatius Donnelly.

Donnelly and the Etruscans

Ignatius Donnelly was a congressman from Minnesota

from 1863 until 1869. While living in Washington,

D.C., he spent much of his time doing research at the

Library of Congress, and subsequently used the data to

write a book about Atlantis. The book, published in

1882, was titled *Atlantis: the Antediluvian World*. It was a bestseller, reprinted many times since. It sparked an Atlantis craze and even inspired a Mardi Gras theme. Donnelly is considered to be the father of modern Atlantology, and has influenced writers in the genre to the present day. Madame Blavatsky acknowledged Donnelly's influence in her own work.

Donnelly attributed the origins of the Etruscans to Atlantean colonization. He offered as proof the rather questionable evidence that Etruscan bronze artifacts resembled those of Atlantean origin. The problem with this is that there are no such recognized artifacts. Although much of what Donnelly said is unsubstantiated, some of his claims are intriguing. He said the Basques of northern Spain and southwestern France, for example, were of Atlantean heritage because they differed from all of their neighbors in appearance and language, and that they migrated from Atlantis before it disappeared beneath the sea.

Author Frank Joseph, in an essay for *Atlantis Rising* magazine, wrote that the Basque language, Euskera, contains the word *Atlaintika*, which is the name of a drowned kingdom from which their ancestors supposedly arrived in the Bay of Biscay. It is true that the Basque language has no close affinity with any other language. This is also the case for the Etruscan language, which had no roots in any of the other Indo-European languages.

Although Donnelly may have been right regarding the provenance of the Etruscans, there is a better reason for it than he suggested. This can be found in Plato's explanation that Atlantis controlled Europe up to Tyrrhenia.

If that were the case, it seems probable that Atlantis established a permanent settlement in ancient Italy for the purpose of governing. Since the Greeks called the Etruscan people the Tyrrhenoi, their country Tyrrhenia, and the adjacent sea the Tyrrhenian Sea— and with their obvious similarity to the names of Tyrus and Gadirus—it can reasonably be concluded that the Etruscans had their origins in Atlantis. When the unique or anomalous quality of their language is also considered, their Atlantean origin seems even more likely.

The Phoenicians and the Island of Erytheia

Historians believe the Phoenicians were Canaanites, a Semitic-speaking people who migrated from northern Arabia east to the Tigris–Euphrates Valley, as well as west toward the Mediterranean. Their earliest known history dates from about 2800 B.C., but they left no written records of themselves.

The relationship of the Island of Erytheia to the origins of the Phoenicians is important but confusing. This is because the Island of Erytheia shared its name with the Erythraean Sea, which in the past was also the name of several different bodies of water in the Near and Middle East.

In Greek mythology, Erytheia was an island in the far western streams of the River Okeanos, bathed in the red light of the setting sun, along with other mythical realms such as Hesperia, the garden of the gods. Hercules was sent to the Island of Erytheia as one of his twelve labors to fetch the herd of red-skinned cattle from the giant Geryon.

Many Greek writers identified Erytheia with ancient Cádiz and southern Spain. The Erythraean Sea, on the other hand, was once the name of the NW Indian Ocean, including the Gulf of Aden and the Persian Gulf, as well as the Red Sea between Egypt and Saudi Arabia. The modern country of Eritrea was also named after the Erythraean Sea.

One Greek writer, Pseudo-Apollodorus, wrote in *Bibliotheca* that Erytheia was an island, in his time called Gadeira, near Okeanos. This shows a link between Erytheia and Plato's Atlantis, because Plato said Gadeira was an area of Spain named after Atlas's twin brother Gadirus, the ruler of that part of Atlantis nearest to Spain. Strabo, in Geography, also linked Erytheia with Gades, the ancient name for Cádiz, and identified it as a city and island off the coast of Southern Iberia. And finally, Herodotus wrote in his *Histories* that Erytheia was an island near Gadeira outside the Pillars of Hercules.

These Greek writers, and others such as Pliny the Elder, Philostratus, and Stesichorus all support the idea from Greek mythology that Erytheia was located in the far west, at least in relation to Greece. The confusion about the origins of the Phoenicians lies in the disparity between the western location of Erytheia and the eastern location of the Erythraean Sea.

Herodotus wrote in *Histories*, for example, that the Phoenicians came originally from the Erythraean Sea. Pliny the Elder also said in his *Natural History* that Erytheia derived its name from the fact that the Tyrians, the original ancestors of the Carthaginians, came from the Erythaean, or Red Sea. Of particular interest in Pliny's statement is the link between the

Phoenician city of Carthage, the Tyrians, and Erytheia. It could be argued from Pliny that the Tyrians, the ancestors of the Phoenicians, originated in Erytheia—that area of the world in which Cádiz is located, along with its ties to Plato's Atlantis.

That is, in fact, just what Ignatius Donnelly argued in *Atlantis: the Antediluvian World*. According to Donnelly, ancient Persian traditions say that the Phoenicians migrated from the shores of the Erythraean Sea. This has generally been understood to mean from the shores of the Persian Gulf, which lies to the east and south of the eastern Mediterranean, and is in agreement with the accepted historical record. Donnelly argued, however, that the Erythraean Sea was originally adjacent to the mythological Greek Island of Erytheia in the west, and took its name from that island. He further suggested that Erytheia was the starting point of the Phoenicians in their European migrations. When they migrated from the Erythraean Sea, or the Atlantic, said Donnelly, they first gave their name to the city on the coast of Spain, then later to the Persian Gulf.

Tyrus and Tyre

Strabo wrote in Geography that historians before him claimed at least 300 ancient Tyrian cities once existed on the coast of Africa beyond the Pillars of Hercules. He regarded these stories as fictitious, but were they really just fables? If the Tyrians were the ancestors of the Phoenicians, and originated in the Atlantic in the region of Erytheia, it's reasonable to believe they may have established settlements on the coast of Africa outside the Pillars of Hercules.

If Tyrus was a part of Atlantis, as Blavatsky said, and the original home of the Tyrians, and Ezekiel's description of its location was accurate, then only one logical conclusion can be reached: Tyrus and the Phoenician city of Tyre were two different cities, widely separated by time and distance. They were not simply spelling variations of the same city.

If true, then the confusion between the two cities is understandable given the similarities between their names and the circumstances of their locations. The use of the two different spellings throughout the Old Testament must have been due to an error in the compilation of the materials used to write the various accounts of Tyrus and Tyre.

Scholars concede that Ezekiel was compiled from many different source documents, and although it had a primary redactor, it is thought to be the work of more than one author. These authors may have been unaware of the difference between the two cities and combined the various source materials in an attempt to create one coherent story. The result was a conflated and confused account of Atlantis and the Phoenician city that continues to this day.

So, it's likely that survivors from the destruction of Atlantis sailed through the Strait of Gibraltar to the future site of the Phoenician city of Tyre, and named the new city after the old. It's human nature to do such a thing. A study of the map of the eastern United States shows many cities of European origin. New York, to use Ignatius Donnelly's example, along with Plymouth, Greenwich, Dover, Gloucester, and many others, were all named after cities in England. The idea that the Phoenician city of Tyre was named after a city of Atlantis called Tyrus is certainly plausible.

6

Zidon

Bronze sculpture of Neptune, Virginia Beach, Virginia

Neptune was the Roman name for Poseidon

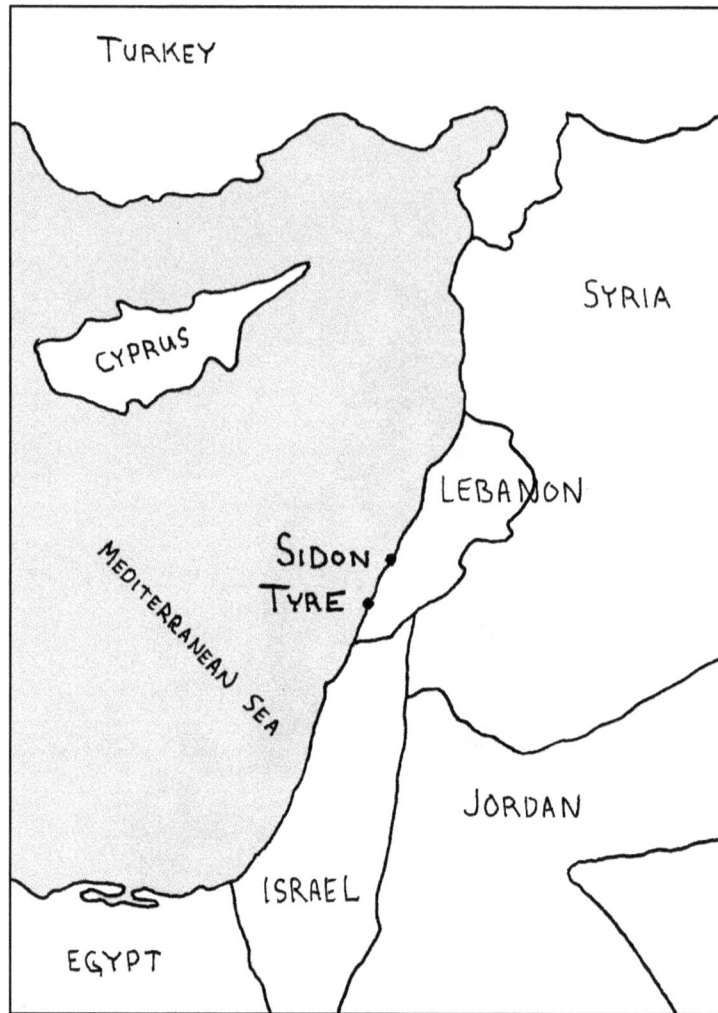

Location of the Phoenician cities Sidon and Tyre

Much of the discussion so far concerning Tyrus and Tyre can also be applied to Zidon, which is assumed to have been the Phoenician city of Sidon—located on the shores of the eastern Mediterranean—and one of the most ancient of the Phoenician city-states.

Sidon lies on the Lebanese coast about twenty-five miles north of Tyre. The modern city is now called Saida, and exists on the ruins of Sidon. The exact antiquity of Sidon isn't known, but, like Tyre, was known to be in existence sometime before 2000 B.C. The city was once a commercial empire, the leader of the Phoenician cities during the second millennium B.C.

After the eleventh century B.C., however, Sidon's influence declined. It became secondary in significance to Tyre.

Zidon is almost always mentioned in the Old Testament in association with Tyrus. This seems reasonable if they were the Phoenician cities of Sidon and Tyre, but it would also be the case if they were both cities of Atlantis.

According to John McKenzie in *Dictionary of the Bible* (1965), Phoenicia was never politically unified. It was made up of independent city-states, the most important being Sidon, Tyre, Byblos, Arwad, and Ugarit. These city-states were usually jealous and hostile to each other, but Ezekiel and Isaiah describe an atmosphere of peaceful cooperation and trade

between Zidon and Tyrus, as might be expected if,

again, they were part of Atlantis.

The inhabitants of Zidon and Arvad were thy mariners:

thy wise men, O Tyrus, that were in thee, were thy pilots.

— Ezekiel 27:8

Be still, ye inhabitants of the isle [Tyre]*;*

Thou whom the merchants of Zidon,

That pass over the sea, have replenished.

— Isaiah 23:2

Another example of the close association between

Tyrus and Zidon occurs at the end of Ezekiel's long

prophecy concerning the destruction of Tyrus. The chapter ends with a prophecy for the destruction of Zidon, as well:

Son of man, set thy face against Zidon, and prophecy against it.

— Ezekiel 28:21

And say, Thus saith the Lord God; Behold, I am against thee. O Zidon; and I will be glorified in the midst of thee: and they shall know that I am the Lord, when I shall have executed judgements in her, and shall be sanctified in her.

— Ezekiel 28:22

History shows that the actual Phoenician cities of Tyre and Sidon were never destroyed, so it's quite possible that Ezekiel's prophecy of destruction was really about two entirely different cities located somewhere else.

Zidon and Poseidon

There is an obvious similarity between the names *Poseidon*, the god Plato said was allotted the island of Atlantis, and *Zidon*. This is also true of the word Sidonian. Although the Greeks gave the Phoenicians their name, which means purple, after the dye they were known for, the Phoenicians referred to themselves as Sidonians. Ancient documents also show that the word Sidonian was used to refer to the inhabitants of the Phoenician coast, which is the strip of land including modern Syria and Lebanon.

Both words possess similarities to Poseidon. Let's read Ezekiel's passage again, aloud:

...Thus saith the Lord God;

*Behold, I am against thee, **O Zidon**...*

— Ezekiel 28:22 [emphasis added]

Newer translations of this passage, such as the one found in the *Revised Standard Version* (1952), have changed the spelling of *Zidon* to *Sidon*, evincing a similar assumption as in the case of Tyrus and Tyre—that Zidon and Sidon were the same city.

Early in its history, Atlantis is believed to have consisted of several islands scattered across the Atlantic Ocean. In the theosophical literature, it is mentioned that before Atlantis finally disappeared, only one island remained, called Poseidonis. Similarly, the psychics Edgar Cayce and Ruth Montgomery, both of whom will be discussed in a later chapter, said the last remaining island of Atlantis was named Poseidia. It is possible that Zidon could have been the biblical name for Poseidonis or Poseidia, and that Edgar Cayce, Ruth Montgomery, and Ezekiel could have been describing the same city.

The Daughter of Zidon

A passage from Isaiah 23 contains an enigmatic

reference to the daughter of Zidon:

> *And he said, Thou* [Tyre] *shalt no more rejoice,*
>
> *O thou oppressed virgin, daughter of Zidon:*
>
> *Arise, pass ye over to Chittim;*
>
> *There also shalt thou have no rest.*

> — Isaiah 23:12

The *New Interpreter's Bible* doesn't specifically

address the phrase "daughter of Zidon," but instead

gives a general explanation for the whole passage,

saying that Zidon will find no rest even if its people flee

to Chittim. The phrase does bring to mind a book by

Taylor Caldwell, *The Romance of Atlantis*, in which the last ruler of Atlantis was described as a woman. The book chronicles the life of the Empress Salustra, presumably the last of the royal bloodline of Poseidon, as she presides over the final days of Atlantis. Taylor Caldwell wrote this remarkable book when she was only twelve years old. Her incredulous grandfather, a book editor, thought it had been plagiarized, and recommended that the manuscript be destroyed. Fortunately, the manuscript survived, and it wasn't until late in Caldwell's life that the book found its way into print. The material in the book is said to have come to Caldwell from a recurring dream, and many believe the story to be evidence of reincarnation. It is doubtful the original meaning of the passage will ever

be known, but maybe the "daughter of Zidon" referred

to someone much like the character of Empress

Salustra in Taylor Caldwell's book.

7

Commerce and the Isles of the Gentiles

A merchant ship of Atlantis

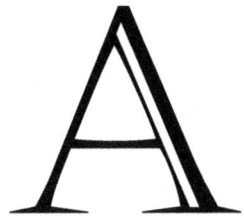

Another similarity between Tyrus and Atlantis lies in the descriptions of the extensive trading and commercial activities engaged in by both cities. Ezekiel in particular gave a long list of the countries and nations that were commercially involved with Tyrus, and also described the multitude of luxury goods that the city received.

Plato gave a brief description of the commercial activity of Atlantis in the Critias.

Because of the extent of their power they received many imports, but for most of their needs the island itself provided.

The long list of Tyrus's trade partners makes up a substantial portion of chapter 27 of Ezekiel. Scholars believe these trading partners, except for Tarshish, were located in the eastern Mediterranean region. The *New Interpreter's Bible* notes that "Ezekiel's account provides many examples of the finest goods the ancient world had to offer," and adds that Ezekiel did this in order to impress the reader with the extent of the wealth and power of Tyrus. The following are typical examples from the list:

The men of Dedan were thy merchants; many isles were the merchandise of thine hand: they brought thee for a present horns of ivory and ebony. — Ezekiel 27:15

The merchants of Sheba and Raamah, they were thy merchants: they occupied in thy fairs with chief of all spices, and with all precious stones, and gold.

— Ezekiel 27: 22

At the beginning of the list, Ezekiel describes the construction of the beautiful "ship of Tyrus." This ship was made of the finest materials available from the ancient Mediterranean world—the oaks of Bashan for the oars, fine linen from Egypt for the sails, cedars from Lebanon for the masts, and so on. According to *Harper's Bible Commentary*, this section of Ezekiel is a metaphor comparing the island of Tyrus to a beautiful, stately ship.

The *New Interpreter's Bible* points out that many scholars find Ezekiel's extensive list of Tyrus's trade partners problematic. They assert that the list doesn't fit the poetic context of Ezekiel, and displays a reworking of an outside source document that was perhaps derived from a government agency or commercial trade house. The implication is that a later redactor extended Ezekiel's original account, perhaps to strengthen what was perceived to be Ezekiel's moral lesson concerning the perils of accumulating enormous wealth. In Isaiah's account of Tyrus, the only places mentioned are Tyrus, Zidon, Chittim, Tarshish, and Chaldea (Assyria). An extensive trade list like the one found in Ezekiel is not included in Isaiah's account.

But just because Tyrus had a successful, far-reaching commercial empire like Atlantis doesn't necessarily prove that it was, in fact, Atlantis. The far-reaching commercial activities of both were certainly very similar, but that characteristic alone doesn't establish a definite link between the two.

Ignatius Donnelly attributed the commercial success of the Phoenicians to their Atlantean heritage. He wrote, "It would appear probable that, while other races represent the conquests or colonizations of Atlantis, the Phoenicians succeeded to their arts, sciences, and especially their commercial supremacy." While Donnelly's suggestion of a link between the

Phoenicians and Atlantis may be true, he doesn't provide any conclusive evidence for his assertion.

There is, however, an interesting feature of Ezekiel's list of trade partners that may point to such a link. It pertains to the extensive involvement of Tyrus with the isles of the Gentiles.

The Isles of the Gentiles

The isles of the Gentiles are first mentioned in Genesis.

> *By these were the isles of the Gentiles divided in their lands; every one after his tongue, after their families, in their nations.* — Genesis 10:5 KJV

The inhabitants of the isles of the Gentiles are the descendants of Noah's eldest son Japheth. They, along with the descendants of Noah's other sons, Shem and Ham, settled the earth after the flood. The genealogy of Noah and his sons is usually referred to as the Table of Nations, and is a history of the world after the deluge.

Rather than a specific tribe or precise geographical location, the names listed in the Table of Nations are usually regarded as geographical groupings of people who share similar languages and live in adjacent lands. For example, the sons of Japheth are thought of as Indo-European, the sons of Shem are the Semites, and the sons of Ham are thought of as African. There are, however, many problems with identifying people and places on the list, and the historical situations the list represents have been the subject of scholarly dispute.

Descendants of Japheth (Isles of the Gentiles)

1st Generation	2nd Generation	Occurrence in Ezekiel
Gomer		38:6
	Ashkenaz	
	Riphath	
	Togarmah	27:14
Magog		38:2
Madai		
Javan		27:13,19
	Elishah	27:7
	Tarshish	27:5
	Kittim	27:6
	Dodanim	27:15
Tubal		27:13
Meshech		27:13
Tiras		

The isles of the Gentiles, or the descendants of Japheth, are generally thought to represent the peoples of Asia Minor and Greece. Tarshish, though, is an exception to this because it has long been associated with a specific place—the classical city of Tartessos, on the mouth of the Guadalquivir, north of Cádiz.

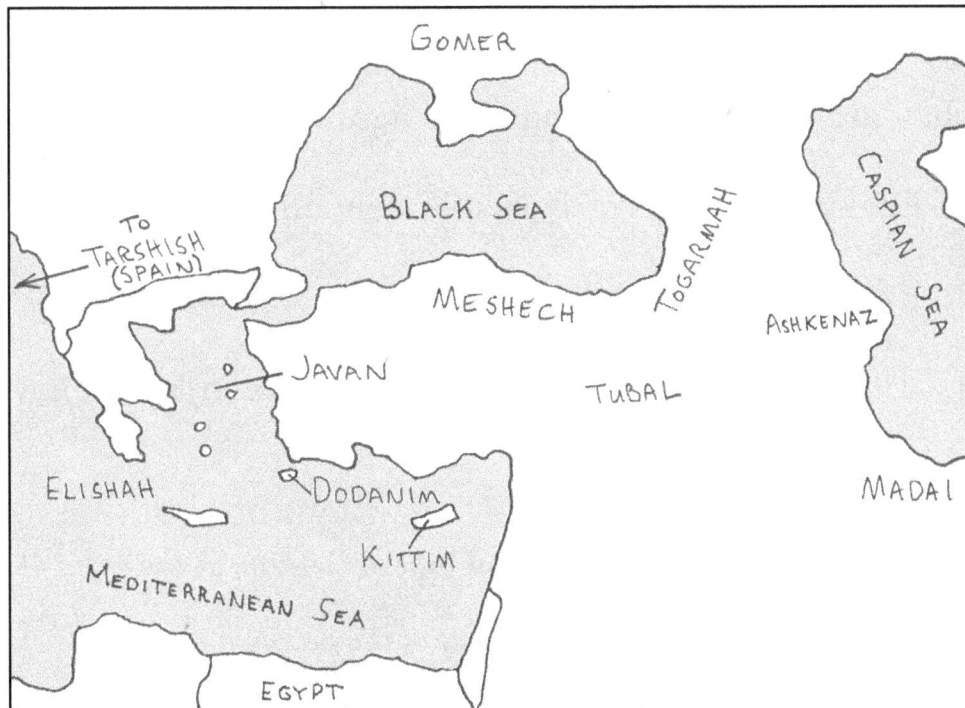

Distribution of the isles of the Gentiles

What is important here is the role the isles of the Gentiles play in Ezekiel. Apart from their first appearance and listing in Genesis, and some scattered references to individual isles in the Old Testament, they are mentioned in the Bible again as a group only in Ezekiel's account of Tyrus. Nearly all are mentioned in Ezekiel's list of trade partners with Tyrus.

Here is the first occurrence of the isles in Ezekiel:

And say unto Tyrus, O thou that art situate at the entry of the sea, which art a merchant of the people for many isles... — Ezekiel 27:3

This passage was discussed earlier because of what it says about the location of Tyrus. What is also noteworthy here is the mention of the isles in conjunction with Tyrus. Although they are not designated as the isles of the Gentiles in this passage, it seems logical to conclude that this is what the author intended because what follows shortly thereafter is the trade list that mentions nearly all of the individual isles of the Gentiles.

Here are the passages from the list that mention the various isles. The first part of the list describes the materials that went into the construction of what the *New Interpreter's Bible* refers to as the "ship of Tyre."

Fine linen with broidered work from Egypt was that which

thou spreadest forth to be thy sail; blue and purple from

the isles of Elishah was that which covered thee.

— Ezekiel 27:7

After the construction of the ship of Tyre is
described, the list of Tyrus's trading partners begins.

Tarshish was thy merchant by reason of the multitude of

all kind of riches; with silver, iron, tin, and lead, they

traded in thy fairs. — Ezekiel 27:12

Javan, Tubal, and Meshech, they were thy merchants: they

traded the persons of men and vessels of brass in thy

market. — Ezekiel 27: 13

They of the house of Togarmah traded in thy fairs with

horses and horsemen and mules.   — Ezekiel 27:14

The list of trading partners mentioning the isles of

the Gentiles concludes with another reference to

Tarshish.

The ships of Tarshish did sing of thee in thy market: and

thou wast replenished, and made very glorious in the

midst of the seas.   — Ezekiel 27: 25

From "isles" to "coastlands"

All of the passages just cited are from the King James

Version of the Bible. Even though the isles of the

Gentiles are described in the KJV as islands throughout the Old Testament, modern reference works such as the *New Interpreter's Bible* have changed the designation of "island" to "coastland peoples" and other similar descriptions. This change has been incorporated into modern translations of the Bible as well.

Genesis 10:5

Version	Text
King James Bible, **1611**	*By these were the isles of the Gentiles divided in their lands.*
New English Bible, **1946**	*From these the peoples of the coasts and islands separated into their own countries.*
New International Version, **1978**	*From these the maritime peoples spread out into their territories.*
New Revised Standard Version, **1989**	*From these the coastland peoples spread.*

The above changes reflect the modern belief that civilization began in the Middle East and spread outward from there. It is believed that these islands and coastal peoples existed for the most part in the eastern Mediterranean and Black Sea or Caucasus regions.

According to modern scholars, the isles of the Gentiles weren't necessarily islands but groups of people living along the coastal areas in these regions. Thus, all modern references to the isles of the Gentiles in Ezekiel's account of Tyrus have been changed.

Compared to the King James Version, which originally described Tyrus as "a merchant of the people for many isles" (Ezekiel 27:3), the *New Revised*

Standard Version (NRSV) says that Tyre was "a merchant of the peoples on many coastlands." Where the KJV describes the "blue and purple from the isles of Elishah" (Ezekiel 27:7), the NRSV says the "blue and purple from the coasts of Elishah."

These changes have drastically affected the meaning of what Ezekiel wrote. A careful study of the KJV and those passages that refer to the isles of the Gentiles seems to contradict the modern opinion of where they were located.

Just as in the case of Tyrus and Zidon, the KJV seems to say that the isles of the Gentiles were located somewhere outside of the Mediterranean Sea. There are a number of reasons for this, but they stem primarily from the observation that the isles of the Gentiles are almost always associated with Tyrus and Zidon.

Atlantis and the isles of the Gentiles

Several other references to the isles can be found in Ezekiel, and in the Old Testament as well, that mention specific names from the isles of the Gentiles.

And I will send a fire on Magog, and among them that dwell carelessly in the isles: and they shall know I am the Lord. — Ezekiel 39:6

To Tarshish, Pul, and Lud that draw the bow,

To Tubal, and Javan, to the isles afar off,

That have not heard my fame, neither have seen my glory;

And they shall declare my glory among the Gentiles.

— Isaiah 66:19

The kings of Tarshish and of the isles shall bring presents.

— Psalm 72:10

Now, with these examples in mind, a re-examination of Jeremiah 25:22 may shed further light on its interpretation.

And all the kings of Tyrus, and all the kings of Zidon, and the kings of the isles which are beyond the sea.

— Jeremiah 25:22

The isles mentioned by Jeremiah are not specifically identified as the isles of the Gentiles, but when the passage is placed side-by-side with the others that refer to the isles, it's easy to conclude that Jeremiah meant the isles of the Gentiles.

Locating the Gentiles, Tyrus, and Zidon in the Atlantic Ocean, beyond the Strait of Gibraltar, contradicts the historical view that the isles of the Gentiles—except for Tarshish—were located in the eastern Mediterranean and Black Sea. Tarshish was thought to be the only one of the isles of the Gentiles located on the Atlantic side of the Strait of Gibraltar.

The Ivory of Chittim

Another of the isles of the Gentiles, Chittim, is mentioned in Ezekiel's trade list as a source of ivory for Tyrus. This is somewhat puzzling for a number of reasons, and leads to speculation about a possible link between Chittim and Atlantis.

Chittim, also known as Kittim in Hebrew, is believed to have been the Phoenician colony of Kition, established on the island of Cyprus sometime around 800 B.C. This belief contradicts the text of Ezekiel, which refers to the isles of Chittim. Cyprus, on the other hand, consists of only one island. Ezekiel mentions Chittim near the beginning of the list of Tyrus's trade partners.

Of the oaks of Bashan have they made thine oars; the

company of the Ashurites have made thy benches of ivory,

brought out of the isles of Chittim. — Ezekiel 27:6

Interestingly enough, Cyprus was inhabited by a

species of dwarf elephant up until the late Pleistocene.

This species of elephant was much smaller than its

ancestors, and weighed only about 441 pounds. Dwarf

elephant remains have also been found on other

Mediterranean islands such as Malta, Crete, Sicily,

Sardinia, and the Dodecanese Islands.

Archaeological finds of whole or partial skeletons of

this elephant are very rare. The first recorded finds on

Cyprus were by the paleontologist Dorothea Bates in a cave in 1902. She presented her discovery in a paper to the Royal Society of London in 1903, titled *Preliminary note on the discovery of a pygmy elephant in the Pleistocene of Cyprus.*

What is puzzling about the dwarf elephant is that it became extinct at the end of the Pleistocene, or about 9000 B.C., which was also the approximate date of the beginning of human activity on Cyprus. Scientists believe the early inhabitants of the island hunted the dwarf elephant to extinction. But this is long before the time when the Phoenicians developed into a commercial power in the Mediterranean. The Phoenicians had a monopoly on navigation from approximately 1500 B.C.–700 B.C., some 7,500 years after the extinction of the dwarf elephant. It seems unlikely that Cyprus, or Chittim, could have been a source of ivory during Phoenician times.

The extinction date for the dwarf elephant coincides nicely, however, with Plato's date for the disappearance of Atlantis in approximately 9600 B.C. Plato wrote in the Critias: "Every kind of animal, domesticated and wild, among them numerous elephants," existed on Atlantis at that time. Ezekiel's description of Chittim seems to imply a connection to Atlantis and a location in the Atlantic Ocean, contrary to the accepted historical opinion that it was the island of Cyprus in the eastern Mediterranean.

Isaiah also mentioned Chittim in his account of the destruction of Tyre. This occurs in the passage where the "daughter of Zidon" was advised to "pass over to Chittim" (Isaiah 23:12). It was argued earlier that Zidon was one of the islands of Atlantis. From Isaiah, it can also be concluded that Chittim was in the vicinity of Zidon, and therefore in the Atlantic Ocean.

Tiras

Tiras was one of the isles of the Gentiles listed in Genesis, but it is not mentioned anywhere else in the Bible. In 1838, the German scholar Johann Christian Freidrich Tuch associated Tiras with the Tyrrhenoi of ancient Italy. Other scholars have suggested this connection as well. It should be remembered that the word Tyrrhenoi was the name the Greeks used to refer to the Etruscans, and that Ignatius Donnelly linked the Etruscans to Atlantis. Thus, Tiras can also be linked to Atlantis because of its association with the Etruscans.

Another reference to Tiras can be found in the ancient Jewish work the *Book of Jubilees*, where it is written that the "inheritance of Tiras consisted of four large islands in the ocean." Notice here the use of the word "ocean," as opposed to the "sea" (i.e., the Mediterranean Sea), to describe the location of the islands. The use of the word ocean could mean the

Atlantic Ocean, and show another link between Tiras and Atlantis.

The similarity of the name of Tiras to Tyrus is obvious, suggesting that it was an earlier form of the name. If true, then it can be argued that Tyrus was one of the isles of the Gentiles. Although this is not mentioned in the *Dictionary of the Bible*, the author does point out the remarkable absence of the Phoenician city of Tyre from the Table of Nations in Genesis 10, and adds that the list must have been written before Tyre rose to prominence.

Antillia and the Isles of the Gentiles

The word *Antillia* was the name given by medieval cartographers to an island in the Atlantic Ocean described as lying at various points between Europe and America.

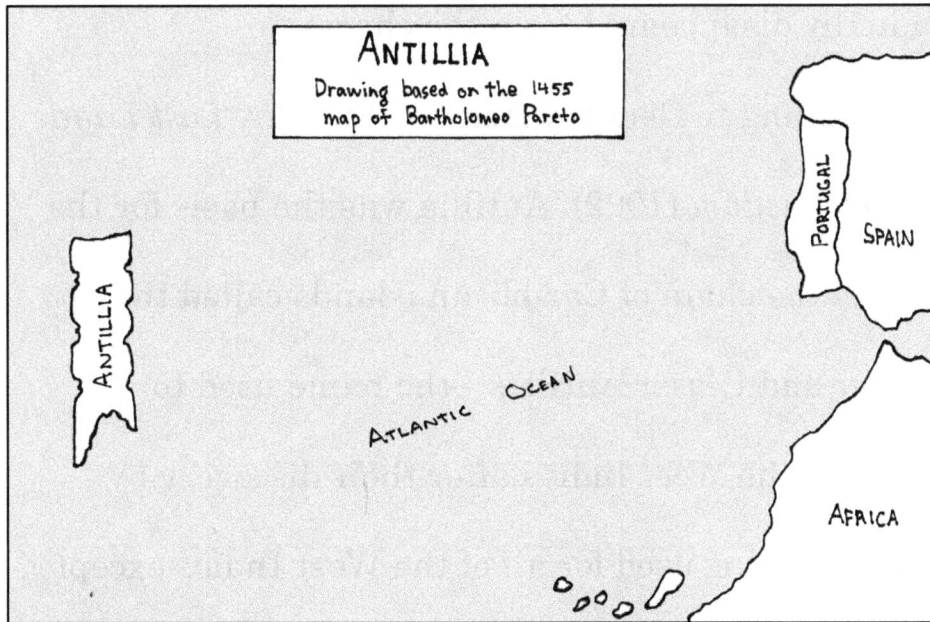

ANTILLIA
Drawing based on the 1455
map of Bartholomeo Pareto

ANTILLIA

PORTUGAL

SPAIN

ATLANTIC OCEAN

AFRICA

Usually indicated as lying far to the west of Spain and Portugal, it first appeared in the 1424 portolan chart of Zuane Pizzigano, and afterward appeared in many nautical charts of the fifteenth century. As the Atlantic Ocean began to be better explored, Antillia gradually disappeared from the charts.

According to Geoffrey Ashe in *Atlantis, Lost Lands, Ancient Wisdom* (1992), Antillia was the basis for the name of the chain of Caribbean islands called the Greater and Lesser Antilles—the name used to designate the West Indies after their discovery by Columbus. It is used for all of the West Indies except for the Bahamas. The islands making up the Greater Antilles consist of Cuba, Hispaniola (Haiti and the

Dominican Republic), Jamaica, and Puerto Rico. The Lesser Antilles are composed of a large string of smaller islands lying to the south.

Several theories have been proposed for the origin of Antillia. One is that Spanish Christians fled by ship after the Muslim conquest of Spain and landed by chance on an island where they founded settlements. Another says the term Antillia may be derived from the Portuguese *ante ilha*, or "facing island," in the belief that the island was situated directly opposite Portugal in the Atlantic Ocean.

Others believe Antillia was a reference to Plato's Atlantis. The resemblance between the two names is obvious. Lewis Spence, in *The History of Atlantis*, differentiates between Atlantis and Antillia, placing Atlantis closer to the Straits of Gibraltar, and Antillia further to the west, where it still remains fragmentally as the West Indies. Spence agreed with Plato's date for the destruction of Atlantis in about 10000 B.C., but said Antillia survived, in part, to the present day.

There is a notable resemblance between the word Gentile and the words Antillia and Antilles. The word Gentile first occurs in the Bible in Genesis 10:5 in the phrase "the isles of the Gentiles." This is the only appearance of the word in the first five books of the Old Testament (the Pentateuch). The phrase "isles of the Gentiles" is a direct translation of *insulae gentium* from St. Jerome's Latin Vulgate, which in turn is a translation of the Hebrew *goyim*. The Latin root of Gentile is gens, meaning family or people, and modern translations of the Bible use the phrase "coastland peoples" instead of "isles of the Gentiles." This change obviously reflects modern thought about the locations of these different peoples, but it is still curious as to why St. Jerome originally described them as islands. The *Book of Jubilees*, as mentioned earlier in the case of Tiras, very clearly describes the isles of the Gentiles as a number of large islands, plus a large land in the north. Did St. Jerome know something that modern scholars have overlooked or ignored?

It is difficult to make a definite connection between the words Antillia and Gentile. A case can be made, however, that the isles of the Gentiles were associated with the islands of Atlantis, because Plato wrote that Atlantis governed not only its own territories but many other islands in the ocean as well. In the Timaeus, he made a careful distinction between the Mediterranean Sea and the real ocean outside the Strait of Gibraltar, so we can assume he meant the Atlantic Ocean. Along with the close association in the KJV between Tyrus and the isles of the Gentiles, it can be argued that they were the same as Plato's "many other islands."

Scholars generally agree that the isles of the Gentiles were located in the eastern Mediterranean region. In addition to the belief that the island of Cyprus was once Chittim, the island of Rhodes is thought to have been Dodanim, and Greece was Elishah.

If true, however, that the isles of the Gentiles were originally located in the Atlantic Ocean, how to explain their accepted Mediterranean locations? The answer is the same as that proposed for Tyrus and Tyre, and Zidon and Sidon: the Mediterranean locations for the isles of the Gentiles were probably later settlements of displaced people fleeing from Atlantis and its territories. Plato wrote that Atlantis sank into the sea in a single day and night, but there was probably plenty of advance warning for the impending catastrophe. The diaspora of the inhabitants of Atlantis and its isles meant that those living nearest the Strait of Gibraltar sailed through it and traveled to the

eastern Mediterranean in order to escape destruction,

carrying with them their culture and the names of

their old homes.

8

Arrogance and War

Warships of Atlantis entering the Mediterranean

Both Tyrus and Atlantis were described as warlike and arrogant. Plato, in the Critias, provided a particularly detailed account of the military makeup and organization of Atlantis. Although Ezekiel does not portray Tyrus quite as warlike as Plato portrayed Atlantis, his description of the city's military preparedness seems somewhat at odds with the historical record of the Phoenician city of Tyre, if that is what Ezekiel's account of Tyrus is really about.

In the Timaeus, Plato recounts what Solon, a relation of Critias's great-grandfather, learned from Egyptian priests about Atlantis and its warlike intentions against the city of Athens.

Our records tell how your city checked a great power which

arrogantly advanced from its base in the Atlantic Ocean to

attack the cities of Europe and Asia.

This dynasty, gathering its whole power together,

attempted to enslave, at a single stroke, your country and

ours and all the territory within the strait.

According to Plato, an Athenian-led alliance of the

Greeks defeated Atlantis and successfully halted its

attempt to conquer the territories in the

Mediterranean. These events are said to have taken

place 9,000 years before Plato's time, but scholars

dismiss this as nonsense. There is no historical or

archaeological evidence of a Greek civilization at that

time. It can be argued, however, that with the destruction of Atlantis, and a subsequent worldwide cataclysm, any such records of an ancient civilization within the Mediterranean Basin would also have been destroyed.

Plato's lengthy and very specific description of the size and makeup of the military of Atlantis can be found in the Critias. He explained that each allotment of land on the island of Atlantis was required to provide men, chariots, and equipment for military service. Since there was a total of 60,000 allotments in each of the ten regions of Atlantis, a nearly unlimited supply of manpower was available for this purpose.

Later, at the end of the Critias, Plato writes that while "at the height of their fame and fortune" the Atlanteans' "pursuit of unbridled ambition and power" led Zeus to punish them "and reduce them to order by discipline."

Ezekiel's "all thy men of war"

Ezekiel prophesizes at length about the judgment and fall of Tyrus, and describes the arrogance, aggression, and warlike tendencies of its inhabitants that would ultimately draw God's wrath upon them.

Therefore thus saith the Lord God; Behold, I am against

thee, O Tyrus, and will cause many nations to come up

against thee, as the sea causeth his waves to come up.

— Ezekiel 26:3

Thy riches, and thy fairs, thy merchandise, thy mariners,

and thy pilots, thy caulkers, and the occupiers of thy

merchandise, and all thy men of war, that are in thee, and

in all thy company which is in the midst of thee, shall fall

into the midst of the seas in the day of thy ruin.

— Ezekiel 27:27

The phrase "all thy men of war" portrays Tyrus as a

warlike city or nation with a sizeable military

component. This contradicts what is known about the

Phoenicians, who preferred trade and profit over war. Ezekiel's description of Tyrus in this instance seems more appropriate for Atlantis than as a description of the Phoenician city of Tyre. The passage implies that the city was more warlike than the historical record would indicate. A primarily mercantilist and commercial-minded nation such as the Phoenician city of Tyre would not maintain a large military force.

A warlike and corrupt Tyrus is further suggested by Ezekiel.

They of Persia and of Lud were in thine army, thy men of war: they hanged the shield and helmet in thee; they set forth thy comeliness.

— Ezekiel 27:10

The men of Arvad with thine army were upon thy walls round about, and the Gammadins were in thy towers: they hanged their shields upon thy walls round about; they have made thy beauty perfect.

— Ezekiel 27:11

Thou hast defiled thy sanctuaries by the multitude of thine iniquities, by the iniquity of thy traffick....

— Ezekiel 28:18

These passages from Ezekiel seem rather exaggerated in the message they convey, and it is understandable that scholars regard the story as metaphorical. This is certainly true if it is actually about the Phoenician city of Tyre. However, if these passages are really about Atlantis, then they accurately reflect Plato's depictions of the island as an arrogant, aggressive, corrupt, and warlike nation. In this case, the passages from Ezekiel can be understood as a literal description of Atlantis, rather than as a metaphorical description of the Phoenician city of Tyre.

Nebuchadnezzar's siege of Tyrus

There is one problem in particular with Blavatsky's claim that Ezekiel's account of Tyrus was really about Atlantis, and it is the description of the attack upon Tyrus by the Babylonian King Nebuchadnezzar.

For thus saith the Lord God; Behold, I will bring upon

Tyrus Nebuchadnezzar king of Babylon, a king of kings,

from the north, with horses, and with chariots, and with

horsemen, and companies, and much people.

— Ezekiel 26:7

The problem this poses is that Nebuchadnezzar attacked the Phoenician city of Tyre in 605 B.C., and held it under siege until 562 B.C. His campaign was unsuccessful, however, and Tyre remained unconquered. Although subdued, the city didn't fall until Alexander the Great besieged and conquered it in 332 B.C. If Blavatsky was right about Tyrus, Ezekiel's account of Nebuchadnezzar's war on Tyre must be explained.

The answer goes back once again to the author or authors who wrote and compiled the book of Ezekiel. If Tyrus was a city of Atlantis, then Ezekiel must have been written at a much earlier time than the occurrence of the siege of the Phoenician city of Tyre. If so, then Ezekiel's original account would necessarily have been shorter than what we know today, and augmented at a later time with material about the siege by authors living during or after the time of the Phoenician city's prominence.

9

The Crystalli Horribilis

The Terrible Crystal

This rather ominous Latin version of "the terrible crystal"—from Ezekiel—can be found in the Vulgate Bible. Eusebius Sophronius Hieronymus, later canonized Saint Jerome by the Catholic Church, began the authorized Latin translation of the Bible in A.D. 382 under a commission from Pope Damasus I. His translation eventually became the official Latin version of the Bible for the Roman Catholic Church.

The English version appears in the King James Bible.

And the likeness of the firmament upon the heads of the living creature was as the color of the terrible crystal, stretched forth over their heads above. — Ezekiel 1:22

No explanation for this mysterious phrase can be found in the *New Interpreter's Bible*. Maybe the phrase was too obscure to warrant a mention. *Harper's Bible Commentary* describes the text of Ezekiel in which the phrase occurs as "occasionally repetitious and contains what appear to be dislocated phrases. For this reason, scholars assume that the account has been secondarily elaborated, either by Ezekiel himself or by later editors."

The word "terrible" has disappeared from many modern translations of the Bible. In the *Revised Standard Version* (1952), for instance, the reference has become "shining like crystal." In the *New International Version* (1978), the phrase has been changed to "sparkling like ice," which completely eliminates the ominous character of the original. The *New English Bible* (1946) takes things even further, changing the KJV's original "color of the terrible crystal" to the wholly benign "glittering like a sheet of ice."

It seems unwise to alter obscure passages in this manner. It would be better to leave them as they are, in the hope that future research will yield their true meaning. Since the NIB offers no explanation for the "terrible crystal," it's necessary to look elsewhere for an explanation. A good place to start is in the material from psychics Edgar Cayce and Ruth Montgomery. They both had much to say about Atlantis, as well as the important role the terrible crystal played there. Of course, sources like these are regarded by academics as questionable at best, and certainly unverifiable. If the contributors to the *New Interpreter's Bible* were even aware of the work of Edgar Cayce and Ruth Montgomery, it seems unlikely they would have afforded any credibility to such material.

Edgar Cayce

Edgar Cayce (1877–1945) is probably the best-known and most-documented psychic of modern times. He was able to enter into a self-induced trance from which he could answer questions on a variety of subjects.

Edgar Cayce

The information is said to have come from the Akashic record, a cosmic or universal database. The answers to the questions he was asked while in trance—readings, as they were called—were given to thousands of people, mostly for the purposes of medical diagnosis and treatment. Over the course of his lifetime, Edgar Cayce gave more than 14,000 readings. The medical advice from the readings was notable because of the accuracy of Cayce's diagnoses, and because they were often given without benefit of seeing the actual patient. His recommendations often involved unusual or natural remedies. Because of this, he has often been called the father of holistic medicine. All of the readings were transcribed and catalogued and have been the subject of many studies and books.

Not all of Cayce's readings were given to address a person's medical problems. Many simply sought advice on personal problems. These *life readings*, as they were called, often gave information about a subject's past life and explained how problems from that life affected the subject's present life.

Over 700 of Cayce's readings were related to Atlantis. Some of the information came about in response to a direct question about Atlantis, but most of the material about Atlantis came through secondarily to the intended reason for that particular reading. What is astonishing about the readings is their internal consistency. Although they were given to many different people, for various reasons, over a period of many years, as a whole, they paint a very consistent and detailed picture of Atlantis. One life reading in particular implied that a certain aspect of Atlantis was mentioned in the Old Testament. This reading described the subject's past life on Atlantis, and said that he was involved in the means of communication and transportation of that time.

Before that we find the entity was in the Atlantean land, during those periods particularly when there was the exodus from Atlantis owing to those activities which were bringing about the destructive forces. There we find the entity was among those who were not only in what is now known as the Yucatan land, but also the Pyrenees and the

Egyptian. For the manners of transportation, the manners

of communications through the airships of that period

were such as Ezekiel described of a much later date.

— Edgar Cayce Reading 1859–1

This reading refers to the first chapter of Ezekiel, in which the prophet described his encounter with a fiery whirlwind, from which emerged the flying throne of the Lord. Cayce said that what Ezekiel described here was an encounter with an Atlantean aircraft, and based on his other readings, with the implication that it was powered by the terrible crystal.

Cayce's explanation sounds like science fiction, of course, but Ezekiel's experience may have been a vision of the remote past from the time of Atlantis, something akin to one of Cayce's trances. What if it wasn't a vision, however, but an actual encounter with an Atlantean aircraft? Based on Plato's date for the destruction of Atlantis, this would necessarily imply that Ezekiel's vision is much older than the accepted timeframe in which Ezekiel is supposed to have actually lived.

Whether Ezekiel had a vision or an actual encounter, another indication of the passage's antiquity is the manner in which he used the phrase the "terrible crystal." It seems to be almost a casual reference, something of general knowledge to the reader, and in no need of further explanation. If so, the meaning has long since been forgotten, and the phrase has become nothing more than an obscure reference.

Ruth Montgomery

Ruth Montgomery (1912–2001) was another well-known psychic in the tradition of Edgar Cayce. She was a professional journalist by trade and spent much of her career around Washington, D.C. She developed an interest in psychic phenomena later in life and wrote many books about her experiences in this area. Unlike Edgar Cayce, who received information while in a trance, Montgomery's books were channeled from her spirit guides through the method of automatic writing. One way in which she differed dramatically from Cayce was her theory of walk-ins. This is when an individual soul can leave a hurt or troubled body and be replaced with a new soul who takes over the body and then carries out its own life tasks. This process seems to result in marked personality changes, according to people familiar with the person involved.

Like Cayce, Montgomery had much to say about Atlantis. Although her material is very similar to Cayce's in this regard, she did provide some extra details about Atlantis and the terrible crystal.

Crystal Power in Atlantis

Cayce and Montgomery said the crystal was used as an important source of energy in Atlantis. It was used to power ships, aircraft, and other forms of transportation. It was also used as a means of communication in those times. How all of this was achieved, of course, is unknown. Apparently, the crystal was carefully shaped into many facets in order to reflect solar energy. It became activated when exposed to the rays of the sun. The energy was then beamed to wherever it was needed. In the present day, the piezoelectric properties of quartz crystals are understood and utilized in various industries such as audio and electronics. The knowledge of the use of quartz crystals as a means to power heavy machinery, however, is unknown and seems rather farfetched.

During the vision sequence at the beginning of Ezekiel, the prophet appears to suggest that the flying throne, or aircraft, was powered by the terrible crystal. Ezekiel first described the sky above the throne as the "color of the terrible crystal," elaborating on this point in a later passage.

And I saw as the color of amber; as the appearance of fire round about within it, from the appearance of his loins even upward, and from the appearance of his loins even downward, I saw as it were the appearance of fire, and it had brightness round about. — Ezekiel 1:27

It is easy to imagine that Ezekiel is describing an activated terrible crystal casting its amber light upon the Lord.

Ezekiel's "Stones of Fire"

In the Cayce readings, the crystal was referred to by several different names—first as the Whitestone, or Tuaoi Stone, and eventually as the Firestone. Interestingly enough, Ezekiel uses the similar-sounding phrase "stones of fire" in a passage about the prince of Tyrus.

Thou art the anointed cherub that covereth; and I have set

thee so: thou wast upon the holy mountain of God; thou

hast walked up and down in the midst of the stones of fire.

— Ezekiel 28:14

The *New Interpreter's Bible* does not mention the possibility of Ezekiel's stones of fire being the same thing as Cayce's Firestone. The NIB regards the phrase as obscure, and scholars have offered several different explanations for it. One sees it as an ancient mythological reference to Baal's abode on Mt. Zaphon, which was constructed of precious metals and jewels fused together by a fire within the building. Another equates it with the coals of fire found on the temple altar and representative of Yahweh's presence. This last explanation can be found in the *New Jerusalem Bible* (1966), in which the KJV's "stones of fire" has been changed to "red-hot coals."

Several scholars have suggested that Ezekiel's stones of fire should be translated as "fire stones," and that the phrase refers to a hedge of precious stones mentioned in an earlier passage. Their opinion is based on a translation of the Akkadian word *aban*.

Akkadian was an ancient Semitic language spoken in Mesopotamia from approximately 2000 B.C. until 500 B.C. It was named after the city of Akkad, located on the Euphrates River and the center of a world kingdom prior to 2000 B.C. Akkadian was eventually supplanted by Aramaic, but a knowledge of the language is important for a better understanding of the Hebrew in which the rabbinical or Masoretic Text of the Old Testament was originally written.

The NIB concludes that none of these explanations for Ezekiel's stones of fire are convincing, and it questions whether Ezekiel invented the phrase or instead simply borrowed it from ancient mythic imagery. What is important here, though, is the suggestion that Ezekiel's "stones of fire" is more accurately translated as "fire stones," because this is nearly identical to Edgar Cayce's Firestone, the word from his readings used to describe the terrible crystal of Atlantis.

Ancient Weapon of Mass Destruction

Cayce and Montgomery each described how the powerful forces generated by the crystal came to be used for the purposes of warfare and weaponry, and how the misuse of the crystal played an important role in the ultimate destruction of Atlantis. Although Plato doesn't mention any such crystal in the Timaeus or Critias, let alone as the cause of the destruction of Atlantis, Ezekiel does seem to imply that the stones of fire—or fire stones—were associated with the destruction of Tyrus.

10

The Destruction of Atlantis and Tyrus

The wrath of Poseidon

There are many similarities between Plato's description of the destruction of Atlantis and Ezekiel's prophecy of the destruction of Tyrus. There are so many, in fact, that it's difficult to believe they weren't describing the same event. Although Plato said Atlantis disappeared almost 12,000 years ago, or approximately in 9600 B.C., and Ezekiel's account was a prophecy of a future event, the two stories parallel each other to a remarkable extent. Here's how Plato described the end of Atlantis in the Timaeus:

> *At a later time there were dreadful earthquakes and floods of extraordinary violence, and in a single dreadful day*

and night all your fighting men were swallowed up by the earth, and the island of Atlantis was similarly swallowed up by the sea and vanished; this is why the sea in that area is to this day impassable to navigation, which is hindered by mud just below the surface, the remains of the sunken island.

And from the Critias:

Atlantis was an island larger than Libya and Asia put together, though it was subsequently overwhelmed by earthquakes and is the source of the impenetrable mud which prevents the free passage of those who sail out of the straits into the open sea.

According to Plato, unbridled ambition and the pursuit of power by the people of Atlantis led to the loss of their morality and their ultimate destruction. As a result, the god Zeus "decided to punish them upon seeing their wretched state of degradation, and consequently reduce them to order by discipline." The dialogue ends abruptly at this point, unfinished, and the reader is left to wonder about the nature of the punishment Zeus had in mind.

Ezekiel repeatedly described the watery destruction of Tyrus in very much the same way as Plato described the destruction of Atlantis. He wrote that the Lord would bring up the deep upon Tyrus, and that great waters would cover it; that Tyrus would fall into the midst of the seas in the day of its ruin; and that it would be broken by the seas in the depths of the waters.

One passage is notable for its relevance to the present day:

> *I will make thee a terror, and thou shallt be no more: though thou be sought for, yet shall thou never be found again, saith the Lord God.* — Ezekiel 26:21

If Ezekiel's prophecy about Tyrus is really about Atlantis, then this passage certainly describes the widespread interest in the subject. If the thousands of books written on Atlantis and the many efforts to find physical evidence of it are any indication, then it has been sought for, and continues to be sought for, right up to this very day.

The Crystal and the Destruction of Atlantis

According to Edgar Cayce, warring factions in Atlantis set off an event that resulted in the final destruction and sinking of the island in approximately 10000 B.C. The two factions consisted of what Cayce called the Sons of the Law of One, on the one hand, and the Sons of Belial, on the other. The first group was made up of those who followed God, while the second group consisted of those who put themselves above God.

Cayce described Atlantis as a warlike and decadent society. Their use of the terrible crystal, which served as a power source for their civilization, extended to destructive ends several times throughout their history. When the crystal was unintentionally tuned to a frequency that was too high, Cayce said, it resulted in widespread destruction of the land.

There were two previous partial destructions of Atlantis before the third and final one, according to Cayce. The first happened in 50722 B.C., and the second one in about 22000 B.C., which resulted in Noah's Flood. Both of the earlier destructions resulted from misuse of the crystal. Although Cayce didn't give the specific reason for the final destruction and disappearance of Atlantis, his description sounds eerily similar to the controversies surrounding the use of nuclear energy in the present day. He was particularly prescient when we consider that he died before the use of the atomic bomb in World War II and the advent of the modern nuclear era.

Ruth Montgomery, in her book *The World Before*, described the destructive role of the crystal much as Cayce did, but in more detail. She said that the final destruction of Atlantis was due to the crystal's power being accidentally increased, causing a build-up of pressure in gas pockets beneath the Atlantic Ridge. The pressure resulted in massive volcanic eruptions, after which Atlantis sank into the sea, accompanied by widespread flooding.

According to Montgomery, the catastrophe occurred about twelve thousand years ago, or 10000 B.C., and was pretty much a repeat of the events that caused Noah's Flood at a much earlier time. Noah's Flood, according to Montgomery, was the result of aiming the rays of the crystal through the earth in an attempt to destroy what is now China. The attempt failed, but it resulted in the Deluge described in Genesis, along with widespread devastation throughout the world.

CROSS SECTION OF THE
EARTH VIEWED FROM
THE NORTH POLE

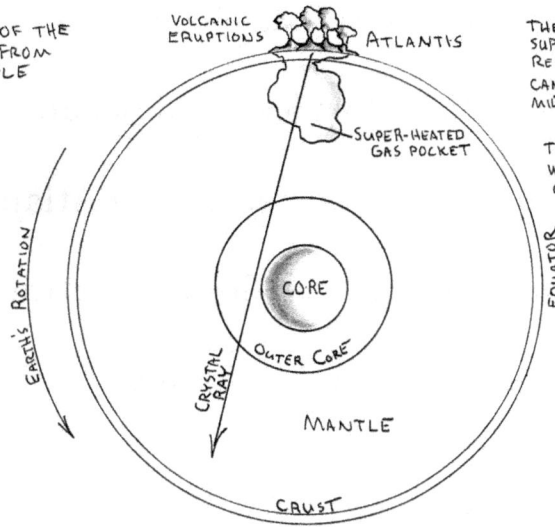

VOLCANIC
ERUPTIONS

ATLANTIS

SUPER-HEATED
GAS POCKET

EARTH'S ROTATION

CRYSTAL RAY

CORE

OUTER CORE

MANTLE

EQUATOR

CRUST

THE PRESSURE FROM THE
SUPER-HEATED GAS POCKET
RESULTED IN MASSIVE VOL-
CANIC ERUPTIONS ALONG THE
MID-ATLANTIC RIDGE.

THE GAS POCKET WAS FORMED
WHEN THE CRYSTAL RAY
OVERHEATED THE EARTH'S
MANTLE.

BEFORE

ANATOMY OF THE CATACLYSM
THAT DESTROYED ATLANTIS CIRCA 9600 B.C.

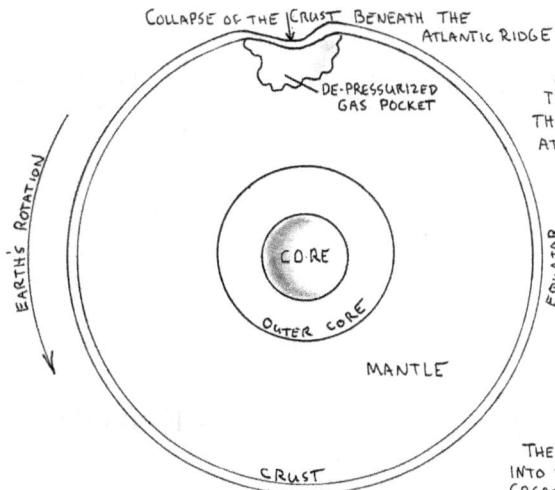

COLLAPSE OF THE CRUST BENEATH THE
ATLANTIC RIDGE

DE-PRESSURIZED
GAS POCKET

EARTH'S ROTATION

CORE

OUTER CORE

MANTLE

EQUATOR

CRUST

AFTER THE PRESSURE IN
THE GAS POCKET WAS RELEASED,
THE EARTH'S CRUST BENEATH
ATLANTIS SUBSIDED INTO
THE MANTLE.

AFTER

THE COLLAPSE OF THE CRUST
INTO THE MANTLE VERY LIKELY
CREATED AN ENORMOUS AMOUNT
OF FRICTION BETWEEN THE TWO
LAYERS, RESULTING IN A CRUSTAL
POLESHIFT.

None of this can be verified, of course, but Ruth Montgomery's mention of widespread flooding, in conjunction with the final destruction of Atlantis, sounds very much like Plato's "floods of extraordinary violence" that accompanied the sinking of Atlantis.

The Stones of Fire & the Destruction of Tyrus

Ezekiel's peculiar phrase "the stones of fire" is also mentioned in conjunction with the destruction of Tyrus.

> *I will destroy thee, O covering cherub, from the midst of the stones of fire.* — Ezekiel 28:16

This passage sounds like Edgar Cayce's and Ruth Montgomery's descriptions of the destruction of Atlantis by the misuse of the terrible crystal. If the stones of fire are the same as Cayce's fire stone, then Ezekiel seems to imply that they will be central to the destruction of Tyrus.

It's informative to read a modern translation of the same passage. The *New International Version* (NIV) reads, "I expelled you, O guardian cherub, from among the fiery stones." Notice how "destroy" has been replaced by "expelled" in the new translation—a much less ominous choice of word.

Volcanic Activity in the Old Testament

Ruth Montgomery said the destruction of Atlantis from misuse of the crystal was accompanied by widespread volcanic activity. Shortly after the destruction of Tyrus from the midst of the stones of fire, Ezekiel seems to say the same thing.

> *I will bring forth a fire from thee, it shall devour thee, and I will bring thee to ashes upon the earth in the sight of all them that behold thee.* — Ezekiel 28:18

Commentary from the *New Interpreter's Bible*, on the other hand, attributes this passage to an ordinary fire capable of destroying a city. Another interprets it as a reference to mid-eastern mythology, in which the resurrection of the god Melqart was celebrated by subjecting the image of the god to ritual cremation.

Zechariah also foretold the same fate for Tyrus.

Behold, the Lord will cast her out, and he will smite her power in the sea; and she shall be devoured with fire.

— Zechariah 9:4

Fire & Brimstone Upon the Isles of the Gentiles

Ezekiel prophesied destruction by fire of several of the isles of the Gentiles, therefore supporting the idea they were originally located near Atlantis, along the Mid-Atlantic Ridge.

And I will send a fire on Magog, and among them that
dwell carelessly in the isles: and they shall know that I am
the Lord. — Ezekiel 39:6

In Ezekiel 38, entitled *Prophecy against Gog*, the prophet explained in greater detail who Magog was, and associated Magog with two more of the isles of the Gentiles.

Son of man, set thy face against Gog, the land of Magog,
the chief prince of Meshech and Tubal.

— Ezekiel 38:2

This passage is followed a short while later with the promise of a rain of fire and brimstone upon Magog.

And I will plead against him with pestilence and with blood; and I will rain upon him, and upon his bands, and upon the many people that are with him, an overflowing rain, and great hailstones, fire, and brimstone.

— Ezekiel 38:22

If the isles of the Gentiles were originally in the vicinity of Atlantis and therefore situated along the Mid-Atlantic Ridge, they could have easily been overwhelmed by volcanic activity. In *When the Earth Nearly Died*, authors D.S. Allan and J.B. Delair cite an enormous amount of scientific data to support their conclusion of a world cataclysm in 9500 B.C. A large part of this destructive event, they say, involved volcanic eruptions. They wrote that the Canaries and

Azores underwent extensive volcanic activity during the late Pleistocene (less than 13,000 years ago), and that the sea floor around the Azores in particular was covered by lava. The authors also mention that the Atlas Mountains once extended into the present Atlantic Ocean at least as far as the Canary Islands. Finally, they cite other geological studies showing that beach sand taken from the bottom of the sea in the Azores proves that the Mid-Atlantic Ridge was above water in relatively recent geological times.

Isaiah also refers to the destruction of mankind by fire after his account of the destruction of Tyre. He tells of an ancient apocalypse and other cataclysmic events that seem to describe the final destruction of Atlantis.

11

The Apocalypse of Isaiah

Destruction by fire

Isaiah 23's account of the destruction of Tyrus is very similar to Ezekiel's, but the two differ in several important ways. Isaiah's account is much shorter than Ezekiel's, and it refers to the city as Tyre rather than Tyrus. Isaiah also refers to the destruction of Tyre as an event of the past, in contrast to Ezekiel, who prophesied the future destruction of Tyrus. This last difference is somewhat puzzling, since Isaiah's ministry began around 742 B.C., while Ezekiel's ministry took place at a later time, around 600 B.C.

The *New Interpreter's Bible* says that attempts have been made to interpret Isaiah 23 as a prophecy of future events, but Isaiah obviously describes the city as already destroyed. This leads to the question of which account is the correct one—Ezekiel's or Isaiah's? If Tyrus was a part of Atlantis, then Isaiah's use of the past tense obviously means his account is the accurate one.

The next chapter of Isaiah describes a devastating apocalypse upon the earth and its inhabitants. Nowhere does Isaiah state this apocalypse resulted from the destruction of Tyre, described in his previous chapter, but the fact that the two chapters follow consecutively seems an unlikely coincidence. This is especially so when it is considered that many features of Isaiah's apocalypse parallel what Plato said about the destruction of Atlantis, and what Ezekiel said

about the destruction of Tyrus.

The Destruction of Humanity by Fire

In chapter 24, Isaiah describes an unknown event in

the remote past in which mankind was destroyed by

fire.

> *Therefore hath the curse devoured the earth, and they that*
>
> *dwell therein are desolate: Therefore the inhabitants of the*
>
> *earth are burned, and few men left.* — Isaiah 24:6

The passage echoes the prophecies of Ezekiel and

Zechariah, both of whom foretold the destruction of

Tyrus by fire. It is also reminiscent of Ezekiel's prophecy in which the Lord "will send a fire on Magog and among them that dwell carelessly in the isles."

The Broken Covenant

According to the *New Interpreter's Bible*, two passages from Isaiah 24 suggest that a universal flood occurred with the apocalypse, but Isaiah provides no details of the event.

> *The earth also is defiled under the inhabitants thereof;*
> *Because they have transgressed the laws, changed the ordinance,*
> *Broken the everlasting covenant.*
>
> — Isaiah 24:5

The broken covenant Isaiah mentions here is obviously the covenant with Noah from Genesis 9:1–17, according to the NIB, where God made an everlasting covenant with all life on earth to never again destroy it with a flood. Isaiah makes another reference to a flood later in the chapter.

For the windows on high are open   — Isaiah 24:18

This again refers to Noah's Flood, in which God opened up the windows of heaven and it rained upon the earth for forty days and forty nights.

The relevance to Atlantis is obvious. It wasn't God, according to Isaiah, who broke the covenant, but man himself, by virtue of defiling the earth and transgressing the laws. This dovetails nicely with the reasons Plato, Edgar Cayce, Ruth Montgomery, and others gave for the destruction of Atlantis—namely, that arrogance, greed, and the lust for power led to the unleashing of forces that led to its ultimate downfall and destruction.

Isaiah's reference to a deluge coincides with other accounts that describe the widespread flooding that accompanied the destruction of Atlantis and Tyrus. Ezekiel, for example, prophesied that great waters would cover Tyrus. Plato wrote of the extraordinary floods that accompanied the sinking of Atlantis, and

Ruth Montgomery also described the occurrence of widespread floods at that time of the destruction.

In *When the Earth Nearly Died*, Allan and Delair make a compelling argument that a worldwide deluge marked the end of the Pleistocene Epoch. They argue that the popular conception of an Ice Age is nothing more than a myth. They say the geological evidence for the Ice Age is much better explained by the movement of large masses of water over the surface of the earth, rather than the movement of glaciers. The Ice Age, in their view, was nothing more than an "icy chimera," a nineteenth century idea erroneously based on the extrapolation of local glacial effects into a worldwide phenomenon.

Another flood reference can be found in Jeremiah.
He mentions an "overflowing flood" in his prophecy
about the Philistines, and later associates it with Tyrus
and Zidon.

Thus saith the Lord;

Behold, waters rise up out of the north,

And shall be an overflowing flood,

And shall overflow the land, and all that is therein:

Then the men shall cry,

And all the inhabitants of the land shall howl.

— Jeremiah 47:2

Because of the day that cometh to spoil all the Philistines,

And to cut off from Tyrus and Zidon every helper that

remaineth...   — Jeremiah 47:4

Pole Shift

Isaiah makes the first of a number of references to
what seems to be a pole shift at the beginning of
chapter 24.

Behold, the Lord maketh the earth empty, and maketh it

waste,

And turneth it upside down, and scattereth abroad the

inhabitants thereof.   — Isaiah 24:1

Although the *New Interpreter's Bible* interprets this as an allusion to a massive earthquake, this verse, as well as the others that come later, seems to describe a much larger catastrophic event than a mere earthquake.

Types of Pole Shifts

The first occurs when the outer crust of the earth slides over the liquid mantle beneath. The axis of the earth itself may or may not change in relation to the orbital plane, but to an observer on earth, the result is the same—the relative position of the outer crust has changed in relation to the earth's rotation about its axis.

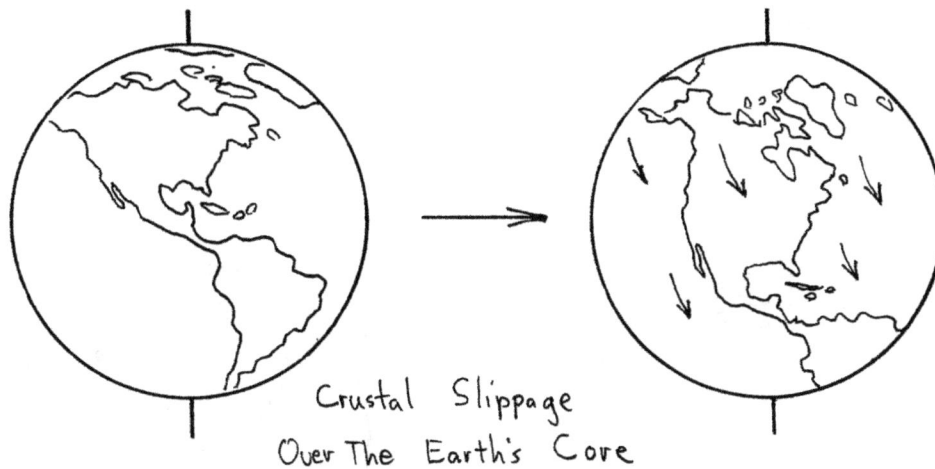

Crustal Slippage
Over The Earth's Core

Another type of pole shift, known as a geomagnetic

pole reversal, is explained in Allan and Delair's *When

the Earth Nearly Died*. The authors say that

geophysicists have long been aware of widespread rock

formations with a reversed polarization. Normally

these rocks began as liquid igneous rocks with a

polarity oriented to the earth's magnetic field as they
cooled.

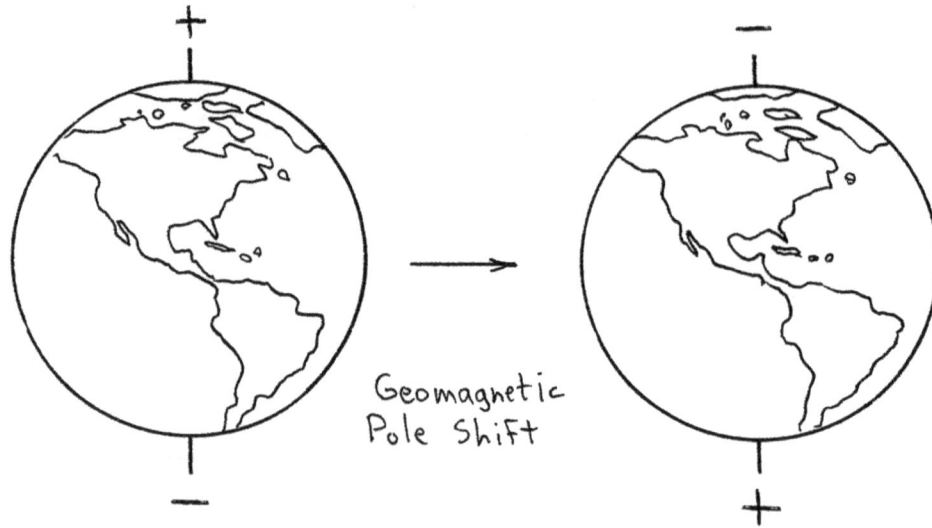

Geomagnetic Pole Shift

The widespread occurrence of rocks with a reversed
polarization means that the earth's magnetic poles
were reversed at some point. The authors cite evidence
that this has happened many times in the past. One
particularly well-marked event of this type, the

Gothenburg Magnetic Reversal, happened in relatively recent geological times—around 9500 B.C. This coincides with Plato's date for the destruction of Atlantis.

The third type of pole shift occurs when the earth topples over or turns upside down in its orbital plane, which is not the same thing as a geomagnetic pole reversal.

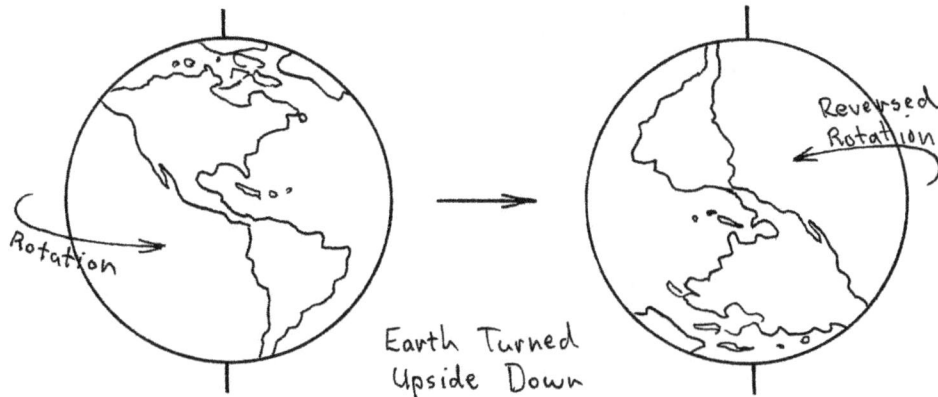

Earth Turned Upside Down

This is the type of pole shift Isaiah seems to describe in the passage quoted above. One effect of this kind of pole shift would be the apparent reversal of the earth's direction of rotation about its axis. The sun would be seen to rise in the west and set in the east.

This may sound counterintuitive, but it can be illustrated with a simple analogy. To an observer stationed above the North Pole, the earth appears to rotate counterclockwise about its axis. This means that to an observer standing on the surface of the earth, the sun rises in the east and sets in the west. If the earth were to topple over, and the observer above the North Pole also moved with the earth as it toppled over, keeping the same position relative to the pole, what would be the result? The earth would still rotate around the axis as before, and the observer above the North Pole, now the South Pole, would still see the earth rotate counterclockwise. But to the observer on the face of the earth, the sun now rises in the west and sets in the east. This is because the earth is now upside-down relative to the orbital plane, and keeps rotating as before, only in the opposite direction.

The reversal of the earth's rotation has been mentioned in several other ancient documents. Immanuel Velikovsky wrote in *Worlds in Collision* that Herodotus described this phenomenon in his own work. While conversing with priests on a trip to Egypt, Herodotus learned that within recorded memory and since Egypt had become a kingdom, "four times in this period the sun rose contrary to his wont; twice he rose where he now sets, and twice he set where he now rises." Velikovsky also mentions, as do authors Allan and Delair, the existence of ancient Egyptian papyri that describe the same catastrophe.

The ancient Egyptian Ipuwer Papyrus tells of a cataclysm that "turned the Earth upside down," while the Ermitage Papyrus, preserved in St. Petersburg, describes a similar ancient world catastrophe.

Different theories have been proposed for the causes of a pole shift. Some argue that only an outside force acting upon the earth, such as a close encounter with a large cosmic object, would have the energy required to cause such a cataclysmic event. Velikovsky, for example, claims that Venus was at one time a wandering planet that nearly collided with the earth on its way to its present orbit around the sun. Allan and Delair argue that a large comet called Phaeton nearly destroyed the earth at the end of the Pleistocene Epoch, and was the source of many ancient disaster legends found throughout the world.

Charles Hapgood, on the other hand, in *Maps of the Ancient Sea Kings*, says the weight of the glaciers during the last Ice Age caused the earth to overbalance and topple over. Albert Einstein himself vouched for the validity of Hapgood's theory on this matter.

The submergence of a continent-sized island such as Atlantis, lying atop the Mid-Atlantic Ridge, would seem capable of precipitating a pole shift. When the crust beneath Atlantis and the underlying Mid-Atlantic Ridge subsided into the molten mantle beneath, a significant amount of friction was likely generated between the two layers.

That portion of the sinking crust, acting as a sort of enormous sea anchor, may have caused the entire earth's crust to skew about in relation to the liquid mantle beneath.

Isaiah continues his description of a pole shift in chapter 24.

The earth shall reel to and fro like a drunkard,

And shall be removed like a cottage.

— Isaiah 24:20

This verse is similar to one from Isaiah 13, "The doom of Babylon."

Therefore I will shake the heavens,

And the earth shall remove out of her place.

— Isaiah 13:13

The *New Interpreter's Bible* once again assumes that these passages describe an earthquake, but qualifies the event by saying that what Isaiah described was no ordinary earthquake. This one was extraordinarily violent and affected the entire world.

Isaiah 24 ends with a description of what an observer might see when looking toward the heavens.

> *Then the moon shall be confounded,*
>
> *And the sun ashamed.*   — Isa 24:23

According to the NIB, this verse shows that "Isaiah's disaster expands to include the 'host of heaven' and the heavenly bodies," and that "God will put an end to all powers, cosmic and earthly." From a scientific standpoint, however, if the earth actually did experience a pole shift that resulted in a movement of the crust, the moon would appear to move erratically in its orbit, perhaps even in a retrograde motion, to someone on earth looking toward the sky during the pole shift.

Although the moon would not actually alter its orbit, it might appear to move or race across the sky in a "confounded way" as the crust moved beneath the observer's feet. If the shift occurred while the observer stood on the nighttime side of the earth, the sun might also be delayed in rising the next day due to the displacement of the crust.

There are several other verses in the Bible that seem to describe a pole shift and the effect it would have on the relative position of the sun.

Behold, I will bring again the shadow of the degrees,

which is gone down in the sun dial of Ahaz, ten degrees

backward. So the sun returned ten degrees, by which

degrees it was gone down.   — Isaiah 38:8

So the sun stood still in the midst of heaven,

And hasted not to go down about a whole day.

                     — Joshua 10:13

And from the Apocrypha:

In his time the sun went backward, and he lengthened the

king's life.   — Ecclesiasticus 48:23

The NIB concludes that Isaiah 24 was inspired by a definite historical event, and although there is no agreement as to what that event was, it must have been far removed from Isaiah's time. Since Isaiah 24 immediately follows his account of the destruction of Tyre—with apocalyptic details that echo what others have said about Atlantis—what other event could Isaiah have described if not the destruction of Atlantis?

12

Lessons from the Past

A dream of Atlantis

It has been argued that the Bible is not a history book and should not be regarded as such. It is primarily a book of faith. However, an awareness of the historical context in which events of the Bible take place is important and is known as Historical Criticism, a branch of the Historical-Critical Method of Bible interpretation. Historical criticism uses such things as archaeology, historical records, and dating technology to confirm the biblical record. In addition, other branches of the Historical-Critical Method use literary analysis, anthropology, and other scientific disciplines to interpret the Bible.

Imagine for a moment that we knew nothing about the Roman Empire other than what is mentioned in the New Testament. How much would we know? The answer is probably very little. Caesar, Pontius Pilate, and Rome are all mentioned, but the full extent and power of the Roman Empire could not be learned from reading the New Testament. Yet, according to the *Dictionary of the Bible*, Israel's history was profoundly influenced by Rome once it became a Roman province, as shown by the events of the Crucifixion—and Christianity spread quickly because of the law and order imposed by the Roman Empire. What we know of ancient Rome comes to us from the many sources that exist outside the Bible.

The Old Testament is full of familiar references to kingdoms such as Israel, Egypt, Babylon, and Assyria, but the locations and origins of other places mentioned in the Old Testament are a mystery. Where exactly, for instance, was Tarshish? Where was the "other side of the flood" mentioned in Joshua 24:14? What about Tyrus and Zidon? Were they really Phoenician cities on the coast of present-day Lebanon? What about the isles of the Gentiles? Were they really coastal cities located in the eastern Mediterranean?

Most historians believe these places were located in the general vicinity of Israel and the eastern Mediterranean. This thinking reflects the modern view that Western civilization started in the Tigris–Euphrates Valley and spread outward from there. But some of the events described in the Old Testament could have taken place under the influence of powerful empires or kingdoms long since forgotten, or relegated to the realm of myth and legend. The idea that a powerful empire existed in the Atlantic Ocean, thousands of years before the emergence of civilization in the Middle East, doesn't fit the modern paradigm. If Atlantis really existed, as Plato said, "beyond the Pillars of Hercules," and it was an aggressive, warlike empire, wouldn't it have profoundly affected the ancient world? Perhaps it did, and the results of its influence remain unrecognized or ignored. Mountains of circumstantial evidence can't prove the existence of Atlantis until incontrovertible proof is discovered—a new type of Rosetta stone, perhaps—showing a clear link between Atlantis and the re-emergence of civilization in the Middle East and the Tigris–Euphrates Valley.

The Roman Empire existed relatively recently in recorded history. Much of what is known about it comes to us from documents and physical evidence that were never lost or destroyed. If Atlantis disappeared in a vast cataclysm around 9600 B.C., or at the end of the Pleistocene Epoch, it's reasonable to assume that most records from such a remote era were destroyed, leaving only fragmentary references to it. Atlantis is regarded as a myth because of the passage of enormous lengths of time. It has been forgotten because of the tragic loss of records from antiquity that were undoubtedly contained in the Library of Alexandria.

Plato explained it best in the Timaeus, in which an Egyptian priest explains to Solon that "writing and the other necessities of civilization have only just been developed when the periodic scourge of the deluge descends, and spares none but the unlettered and uncultured, so that you have to begin again like children, in complete ignorance of what happened in our part of the world or in yours in early times."

In *The World Before*, Ruth Montgomery compares modern America to Atlantis by noting that both are long on invention, mechanical creativity, and a high standard of living, and low on spiritual beliefs and philosophical pursuits. She finds this observation unsettling, but the comparison could be extended to include most of today's industrialized world as well.

A strong case can be made for Madame Blavatsky's assertion that Ezekiel's account of Tyrus was really about Atlantis. But in regard to the lesson Ezekiel sought to impart, it doesn't really matter whether he was speaking about Atlantis or the Phoenician city of Tyre. The lesson is the same in either case: the pride and arrogance that result from placing oneself above God's will, along with the unbridled pursuit of power and wealth, will lead to the destruction of the individual, and ultimately the whole of civilization.

APPENDIX

Selected phrases from the King James Bible and their modern translations.

	KJV King James Version **1611**	GNB Good News Bible **1976**	NIV New International Version **1978**	NJB New Jerusalem Bible **1985**	NRSV New Revised Standard Version **1989**	NET New English Translation **1996**
Ezk 1:22	Color of the terrible crystal	Dome made of dazzling crystal	Sparkling like ice	Glittering like crystal	Shining like crystal	Glittering awesomely like ice
Ezk 27:3	Situate at the entry of the sea	Stands at the edge of the sea	Situated at the gateway of the sea	Enthroned at the gateway of the sea	Sits at the entrance to the sea	Sits at the entrance of the sea
Ezk 27:4	Thy borders are in the midst of the seas	Your home is the sea	Your domain was on the high seas	Your frontiers were far out to sea	Your borders are in the heart of the seas	Your borders are in the heart of the seas
Ezk 28:14	Stones of fire	Sparkling gems	Fiery stones	Amid red hot coals	Stones of fire	Fiery stones
Gen 10:5	By these were the Isles of the Gentiles divided in their lands	The people who live along the coast and on the islands	From these the maritime peoples spread out into their territories	From these came the dispersal to the islands of the nations	From these the coastland peoples spread	From these the coastlands of the nations were separated into their lands

BIBLIOGRAPHY

Allan, D.S. and J.B. Delair. *When the Earth Nearly Died*. Bath, UK: Gateway Books, 995.

Ashe, Geoffrey. *Atlantis, Lost Lands, Ancient Wisdom*. London: Thames and Hudson, Ltd., 1992.

Caldwell, Taylor, and Jess Stearn. *The Romance of Atlantis*. New York: William Morrow & Co., Inc., 1975.

Cayce, Edgar Evans. *Edgar Cayce on Atlantis*. New York: Warner Books, 1968.

Cayce, Edgar Evans, Gail Cayce Schwartzer, and Douglas G. Richards. *Mysteries of Atlantis Revisited*. New York: St. Martin's Paperbacks, 1997.

Donnelly, Ignatius. *Atlantis: the Antediluvian World*. New York: Outlook Book Company, 1985.

Fix, William. *Pyramid Odyssey*. New York: Mayflower Books, 1978.

Freeman, Phillip M. "Ancillary Study: Ancient References to Tartessos." In *Celtic from the West: Alternative Perspectives from Archaeology, Language and Literature*. Edited by Barry Cunliffe and John Koch, 303–334. Oxford, England: Oxbow Books, 2012.

The Good News Bible. New York: American Bible Society, 1976.

Hapgood, Charles. *Maps of the Ancient Sea Kings*. Kempton, IL: Adventures Unlimited Press, 1996.

The Holy Bible, King James Version. London and New York: Collins Clear-Type Press, 1951.

Kenyon, J. Douglas, editor. *Forbidden History: Prehistoric Technologies, Extraterrestrial Intervention, and the Suppressed Origins of Civilization*. Rochester, VT: Bear & Company, 2005.

———. *Forbidden Science: From Ancient Technologies to Free Energy*. Rochester, VT: Bear & Company, 2008.

Little, Gregory L., Lora H. Little, and John Van Auken. *Edgar Cayce's Atlantis*. Virginia Beach, Virginia: A.R.E. Press, 2006.

Montgomery, Ruth. *The World Before*. Greenwich, CT: Fawcett Books, 1976.

———. *Herald of the New Age*. New York: Ballantine Books, 1986.

Plato. *Timaeus and Critias*. London: Penguin Books, 1971. Translation by H.D.P. Lee.

Robinson, Lytle. *Edgar Cayce's Story of the Origin and Destiny of Man*. New York: Berkley Books, 1972.

Sagan, Carl. *Cosmos*. New York: Random House, Inc., 1980.

Smith, Charles Merrill, and James W. Bennett. *How the Bible Was Built*. Cambridge, UK: William B. Eerdmans Publishing Co., 2005.

Spence, Lewis. *The History of Atlantis*. New York: Cosimo, Inc., 1926.

Velikovsky, Immanuel. *Earth In Upheaval*. Garden City, NY: Doubleday & Company, Inc., 1955.

———. *Worlds in Collision*. Garden City, NY: Doubleday & Co., Inc., 1950.

Reference Works

Barnes, Dr. Ian. *The Historical Atlas of the Bible*. London: Chartwell Books, Inc., 2006.

The Bible through the Ages. Pleasantville, NY: Reader's Digest Assoc., Inc., 1989.

Dewey, David. *A User's Guide to Bible Translations – Making the Most of Different Translations*. Downers Grove, IL: Intervarsity Press, 2004.

The Interpreter's Bible. Nashville, TN: Abingdon Press, 1956.